Lecture Notes in Economics and Mathematical Systems

(Vol. 1–15: Lecture Notes in Operations Research and Mathematical Economics, Vol. 16–59: Lecture Notes in Operations Research and Mathematical Systems)

Vol. 1: H. Bühlmann, H. Loeffel, E. Nievergelt, Einführung in die Theorie und Praxis der Entscheidung bei Unsicherheit. 2. Auflage, IV, 125 Seiten. 1969.

Vol. 2: U. N. Bhat, A Study of the Queueing Systems M/G/1 and GI/M/1. VIII, 78 pages. 1968.

Vol. 3: A. Strauss, An Introduction to Optimal Control Theory. Out of print

Vol. 4: Branch and Bound: Eine Einführung. 2., geänderte Auflage. Herausgegeben von F. Weinberg. VII, 174 Seiten. 1973.

Vol. 5: L. P. Hyvärinen, Information Theory for Systems Engineers. VII, 205 pages. 1968.

Vol. 6: H. P. Künzi, O. Müller, E. Nievergelt, Einführungskursus in die dynamische Programmierung. IV, 103 Seiten. 1968.

Vol. 7: W. Popp, Einführung in die Theorie der Lagerhaltung. VI, 173 Seiten. 1968.

Vol. 8: J. Teghem, J. Loris-Teghem, J. P. Lambotte, Modèles d'Attente M/G/1 et GI/M/1 à Arrivées et Services en Groupes. III, 53 pages. 1969.

Vol. 9: E. Schultze, Einführung in die mathematischen Grundlagen der Informationstheorie. VI, 116 Seiten. 1969.

Vol. 10: D. Hochstädter, Stochastische Lagerhaltungsmodelle. VI, 269 Seiten. 1969.

Vol. 11/12: Mathematical Systems Theory and Economics. Edited by H. W. Kuhn and G. P. Szegö. VIII, III, 486 pages. 1969.

Vol. 13: Heuristische Planungsmethoden. Herausgegeben von F. Weinberg und C. A. Zehnder. II, 93 Seiten. 1969.

Vol. 14: Computing Methods in Optimization Problems. V, 191 pages. 1969.

Vol. 15: Economic Models, Estimation and Risk Programming: Essays in Honor of Gerhard Tintner. Edited by K. A. Fox, G. V. L. Narasimham and J. K. Sengupta. VIII, 461 pages. 1969.

Vol. 16: H. P. Künzi und W. Oettli, Nichtlineare Optimierung: Neuere Verfahren, Bibliographie. IV, 180 Seiten. 1969.

Vol. 17: H. Bauer und K. Neumann, Berechnung optimaler Steuerungen, Maximumprinzip und dynamische Optimierung. VIII, 188 Seiten. 1969.

Vol. 18: M. Wolff, Optimale Instandhaltungspolitiken in einfachen Systemen. V, 143 Seiten. 1970.

Vol. 19: L. P. Hyvärinen, Mathematical Modeling for Industrial Processes. VI, 122 pages. 1970.

Vol. 20: G. Uebe, Optimale Fahrpläne. IX, 161 Seiten. 1970.

Vol. 21: Th. M. Liebling, Graphentheorie in Planungs- und Tourenproblemen am Beispiel des städtischen Straßendienstes. IX, 118 Seiten. 1970.

Vol. 22: W. Eichhorn, Theorie der homogenen Produktionsfunktion. VIII, 119 Seiten. 1970.

Vol. 23: A. Ghosal, Some Aspects of Queueing and Storage Systems. IV, 93 pages. 1970.

Vol. 24: G. Feichtinger, Lernprozesse in stochastischen Automaten. V, 66 Seiten. 1970.

Vol. 25: R. Henn und O. Opitz, Konsum- und Produktionstheorie I. II, 124 Seiten. 1970.

Vol. 26: D. Hochstädter und G. Uebe, Ökonometrische Methoden. XII, 250 Seiten. 1970.

Vol. 27: I. H. Mufti, Computational Methods in Optimal Control Problems. IV, 45 pages. 1970.

Vol. 28: Theoretical Approaches to Non-Numerical Problem Solving. Edited by R. B. Banerji and M. D. Mesarovic. VI, 466 pages. 1970.

Vol. 29: S. E. Elmaghraby, Some Network Models in Management Science. III, 176 pages. 1970.

Vol. 30: H. Noltemeier, Sensitivitätsanalyse bei diskreten linearen Optimierungsproblemen. VI, 102 Seiten. 1970.

Vol. 31: M. Kühlmeyer, Die nichtzentrale t-Verteilung. II, 106 Seiten. 1970.

Vol. 32: F. Bartholomes und G. Hotz, Homomorphismen und Reduktionen linearer Sprachen. XII, 143 Seiten. 1970. DM 18,–

Vol. 33: K. Hinderer, Foundations of Non-stationary Dynamic Programming with Discrete Time Parameter. VI, 160 pages. 1970.

Vol. 34: H. Störmer, Semi-Markoff-Prozesse mit endlich vielen Zuständen. Theorie und Anwendungen. VII, 128 Seiten. 1970.

Vol. 35: F. Ferschl, Markovketten. VI, 168 Seiten. 1970.

Vol. 36: M. J. P. Magill, On a General Economic Theory of Motion. VI, 95 pages. 1970.

Vol. 37: H. Müller-Merbach, On Round-Off Errors in Linear Programming. V, 48 pages. 1970.

Vol. 38: Statistische Methoden I. Herausgegeben von E. Walter. VIII, 338 Seiten. 1970.

Vol. 39: Statistische Methoden II. Herausgegeben von E. Walter. IV, 157 Seiten. 1970.

Vol. 40: H. Drygas, The Coordinate-Free Approach to Gauss-Markov Estimation. VIII, 113 pages. 1970.

Vol. 41: U. Ueing, Zwei Lösungsmethoden für nichtkonvexe Programmierungsprobleme. IV, 92 Seiten. 1971.

Vol. 42: A. V. Balakrishnan, Introduction to Optimization Theory in a Hilbert Space. IV, 153 pages. 1971.

Vol. 43: J. A. Morales, Bayesian Full Information Structural Analysis. VI, 154 pages. 1971.

Vol. 44: G. Feichtinger, Stochastische Modelle demographischer Prozesse. IX, 404 Seiten. 1971.

Vol. 45: K. Wendler, Hauptaustauschschritte (Principal Pivoting). II, 64 Seiten. 1971.

Vol. 46: C. Boucher, Leçons sur la théorie des automates mathématiques. VIII, 193 pages. 1971.

Vol. 47: H. A. Nour Eldin, Optimierung linearer Regelsysteme mit quadratischer Zielfunktion. VIII, 163 Seiten. 1971.

Vol. 48: M. Constam, FORTRAN für Anfänger. 2. Auflage. VI, 148 Seiten. 1973.

Vol. 49: Ch. Schneeweiß, Regelungstechnische stochastische Optimierungsverfahren. XI, 254 Seiten. 1971.

Vol. 50: Unternehmensforschung Heute – Übersichtsvorträge der Züricher Tagung von SVOR und DGU, September 1970. Herausgegeben von M. Beckmann. IV, 133 Seiten. 1971.

Vol. 51: Digitale Simulation. Herausgegeben von K. Bauknecht und W. Nef. IV, 207 Seiten. 1971.

Vol. 52: Invariant Imbedding. Proceedings 1970. Edited by R. E. Bellman and E. D. Denman. IV, 148 pages. 1971.

Vol. 53: J. Rosenmüller, Kooperative Spiele und Märkte. III, 152 Seiten. 1971.

Vol. 54: C. C. von Weizsäcker, Steady State Capital Theory. III, 102 pages. 1971.

Vol. 55: P. A. V. B. Swamy, Statistical Inference in Random Coefficient Regression Models. VIII, 209 pages. 1971.

Vol. 56: Mohamed A. El-Hodiri, Constrained Extrema. Introduction to the Differentiable Case with Economic Applications. III, 130 pages. 1971.

Vol. 57: E. Freund, Zeitvariable Mehrgrößensysteme. VIII, 160 Seiten. 1971.

Vol. 58: P. B. Hagelschuer, Theorie der linearen Dekomposition. VII, 191 Seiten. 1971.

continuation on page 169

Lecture Notes in Economics and Mathematical Systems

Managing Editors: M. Beckmann and H. P. Künzi

156

Bertil Näslund

Bo Sellstedt

Neo-Ricardian Theory

With Applications to Some Current Economic Problems

Springer-Verlag

Berlin Heidelberg New York 1978

Authors
Bertil Näslund
Stockholm School of Economics
Box 6501
11383 Stockholm/Sweden

Bo Sellstedt
Department of Business Administration
Stockholm University
10691 Stockholm/Sweden

ISBN-13:978-3-540-08763-2
DOI: 10.1007/978-3-642-46373-0
e-ISBN-13:978-3-642-46373-0

AMS Subject Classifications (1970): 90A15

Printing and binding: Beltz Offsetdruck, Hemsbach/Bergstr.
2142/3140-543210

PREFACE

During the last fifteen years certain very fundamental aspects of economic theory have been very actively debated. The debate has often been called the Cambridge controversy, since the main participants in the debate come from Cambridge, USA and Cambridge, England.

In this debate certain important foundations of economic theory have been criticized which are specifically important for understanding price formation and income distribution. An important starting point has been Sraffa's extensions of Ricardo's theory and one has therefore sometimes called those neo-Ricardians, who have based themselves on Sraffa's work.

We shall deal with some important issues that have been debated, and we shall try to describe a rather complete theoretical construction which will be an alternative to neo-classical theory.

A special problem relates to the title of the book and classification of authors to whom we refer. Perhaps post-Keynesian would sometimes be a better label than neo-Ricardian. We have used the latter name for the book due to its large emphasis on Sraffa's work.

To understand the main parts of the book only basic algebra and calculus are necessary.

The book is divided into two parts, where Part I presents the historical development concentrating on Ricardo and Sraffa. The reason for this is that Sraffa has been of great importance for many aspects of the above mentioned debate and his work builds, as we have pointed out, to a large extent on Ricardo. In the rest of Part I we deal with some major consequences in economics of Sraffa's work and also some ways, in which his theory can be made more complete.

The theories that we describe in Part I are applied to some current economic problems and areas of social concern.

The book has been written as we were teaching a doctoral course in neo-Ricardian economics. We would like to thank the participants in this course for many valuable discussions.

CONTENTS

PART I

THEORY AND DEBATE

CHAPTER 1 WHAT IS NEO-RICARDIAN THEORY?

Economic theory deals with systems, economic systems. Economists therefore meet problems that are common within all analyses of systems, namely

(a) definition of the boundaries of the system (size)

(b) specification of the structure of the system (depth and complexity)

With respect to (a) we know that all economic systems are subsystems of larger systems, e.g. large economic systems or large social and ecological systems. Furthermore, it is necessary to abstract from the real system being studied ((b) above). How the boundaries of the system are set and how its structure is specified naturally depends upon the problem being studied (e.g. macro or micro) but it also depends upon the person studying the problem. Persons observing the same phenomena can represent it with different abstract systems. The difference may depend upon different purposes and different frames of reference.

Since the number of problems that a person can study is very large, a choice is necessary. The problems chosen also depend upon the person's purpose and frame of reference.

Two types of purposes or goals can often be distinguished, namely practical and theoretical ones. To take an example: Marx's main practical purpose was to change the economic structure of society. His theoretical purpose was "to investigate the capitalistic society, its economic and social development and its eventual decline and destruction" (Sommerville, 1972, p. 21). The theoretical purpose was no doubt of less importance than the practical one.

It is clear that if a practical purpose is stressed then the historical conditioning in the choice of problem will be important which naturally is also the case if theoretical purposes are emphasized (it is difficult to imagine a monetary theory without the existence of a monetary economy). However, the important difference is that the specification of the structure of the system will be different (the objectives determine the methods).

It is probably true that most classical economists emphasize institutional circumstances (historical conditioning in the sense mentioned above). In neo-classical theory one has rather tried to arrive at generalizations which should be true in every type of exchange economy (the theoretical objective), e.g. institutions play a much smaller role. [1)]

We also have another form of historic conditioning, namely existing theories, ideas and methods. The development of utility theory was influenced by Bentham's philosophical ideas and the existence of calculus and the methods used by Marx depended upon Hegel's philosophy.

The choice of problem and the structuring of problems are also affected by <u>subjective conditioning</u>; emotions, wishes, opinions. One has a tendency to see what one wants to see.

Two rather different respresentations of a modern economy are

(a) the neo-classical system
(b) the neo-Ricardian system

Between different theoretical systems there might be different kinds of conflicts. One relates to the relation between the real system and the theoretical system; e.g. one theoretical system uses individuals as components while another system uses social classes. This type of difference can be the result of the historical and subjective conditioning mentioned above.

Another difference between theoretical systems can be that concepts which they are using are incompatible. The basic concept in economics has been "value" in one form or another (see Myrdal, 1972, p. 89). One has tried to find a more fundamental value than the exchange value or price; something that can explain price (this does not apply to modern neo-classical theory based on revealed preference), or at least serve as a stable measuring rod for e.g. analysis of exchange value. A theory of value gives inputs, constants, to the system of equations often used by economists to describe an economic system (see Dobb, 1940, p. 5 ff). The conditions which a theory

---

1) Sometimes certain neo-classical economists are accused of having the practical objective of defending existing social conditions and institutions.

of value must fulfill are according to Dobb (1940, p. 9) that

(1) inputs or constants produced by the value theory must be related to something which itself is not value

(2) the theory explains income distribution and price formation (the price of goods)

A theory of value which fulfills condition (1) above can help us if we are interested in causal relations (e.g. what gives a product its value is the amount of labor embodied in it). If one is interested only in predictions then the concept of value is of less importance.

If one tries to briefly summarize the difference between neo-classical and neo-Ricardian theories one can emphasize that they have different theories of value. Neo-classical theory has its roots in utility theory (marginal utility) and neo-Ricardian economics has its roots in labor theory of value.

Different types of differences and conflicts can play different roles during the evolution of a theory. Certain contradictions can be eliminated by integrating two theoretical systems. Other contradictions remain and different theoretical systems develop rather independently of each other.

Other contradictions can have the effect that the existence and acceptance of the system are at stake (cf. Foucault, 1972, p. 175). It is the latter kind of conflict which exists between the neo-Ricardian and neo-classical systems. This probably explains why the debate has been rather agitated.

Starting from their concept of value the neo-Ricardians following Sraffa, have attacked the neo-classical theory of capital and distribution. The neo-classical concept of capital has turned out to be inapplicable for certain kinds of analysis. The debate has been and is theoretical but it has consequences for the answers to many questions that can be asked about several aspects of a capitalistic economy. This probably explains the intensity of the debate. Neo-classical economists have been forced to admit that Sraffa was right but as Joan Robinson (1973, p. 120) has put it:

"He who is convinced against his will
Is of the same opinion still"

Joan Robinson's observation may be explained by a willingness to accept neo-Ricardian criticism of the neo-classical theory, or parts of it, but not to agree with what neo-Ricardians offer in return.

This is specifically so, since the latter build on Ricardo and also on Marx, which means that much is based on a labor theory of value.

We end this chapter by observing that the neo-classical and neo-Ricardain systems are two different representations of a modern economy. Using Nell (1974) we can illustrate the two systems using Figures 1:1 and 1:2.

Some differences between the two theories can be summarized as follows:

(1) Neo-Ricardian theory puts an emphasis on the interdependence of production rather than interdependence in the market

(2) In neo-classical theory substitution is the central mechanism both with respect to supply and demand. This is not the case in neo-Ricardian theory

(3) In neo-classical theory the role of the consumer is central and his goals together with initial endowments determine the prices on the market. In neo-Ricardian theory prices are mainly determined by the conditions of production

(4) In neo-Ricardian economics a distinction is made between workers and capitalists (a vertical analysis to use Robinson's terminology) and from this it follows that an important distinction is made between property and means of production. Neo-classical theory is more atomistic (households, not social classes)

(5) In neo-Ricardian theory the distribution of income is the result of social institutions (e.g. ownership of property) and social relations. In neo-classical theory income distribution is the result of market processes and the effects of the factors of production, labor and capital receive an income corresponding to their marginal productivity

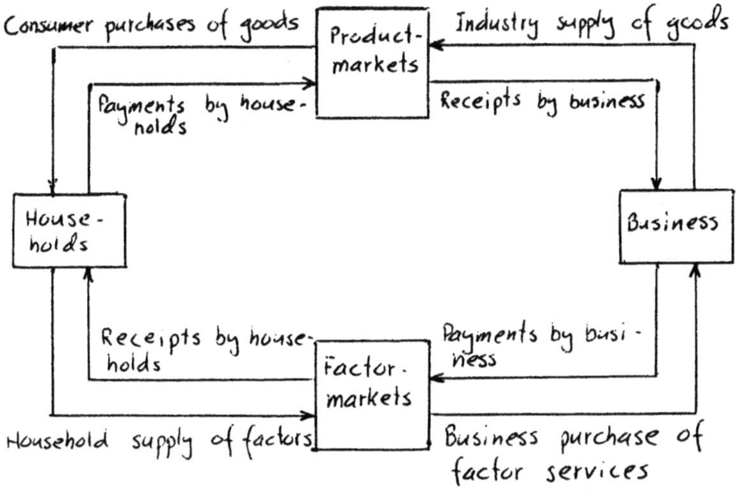

Figure 1:1 The neo-classical view

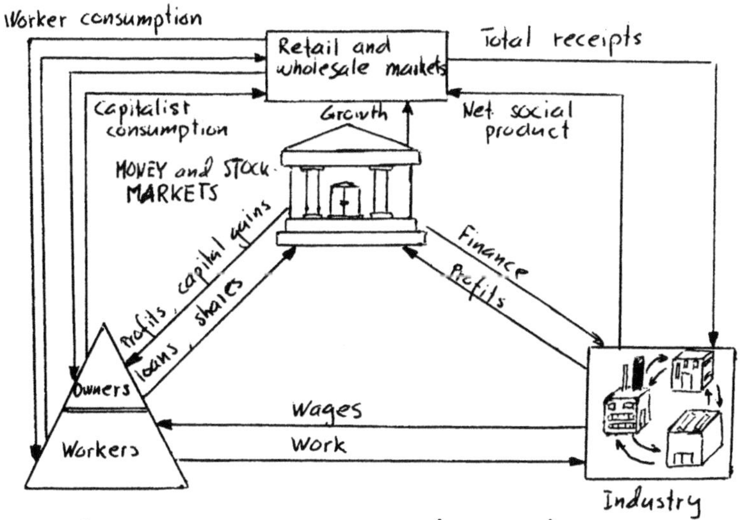

Figure 1:2 The neo-Ricardian and post-Keynesian view

(6) Neo-Ricardian economics has fewer behavioral assumptions.

(7) The theoretical goals are different. In neo-classical theory the prediction of the behavior of the economic agents and the consequences of their behavior are emphasized. In neo-Ricardian economics the purpose is rather to describe how the economic system functions and therefore causal relations are more important.

In this book we are not trying to make a deep analysis of the roots of the two theories. In Chapter 2 we shall, however, follow certain main developments (to a great extent following Dobb, 1973). We hope this will give a deeper understanding of the problems dealt with by neo-Ricardian economics and the methods it uses.

CHAPTER 2 HISTORY EMPHASIZING RICARDO

2.1 The time before Ricardo

One cannot say that a coherent economic theory existed before Ricardo. There existed doctrines and schools of thought which emphasized different aspects of an economy.

These schools concentrated on circumstances that were specific of their time. Mercantilism which blossomed during the 17th and 18th centuries emphasized the importance of foreign trade. This was during the period when the British foreign trade enjoyed strong growth. The mercantilists realized that foreign trade was important for the domestic economy. They knew that a negative trade balance was not good for production at home. If a product is imported then no income is generated for its production. Export, on the other hand, was considered to be equivalent to demand without corresponding supply. Income was earned on goods produced and this income was the source of demand for domestic goods which thereby stimulated the economy. Important economists during that time are David Hume (1711-1776) and James Stuart (1712-1780).

The physiocrats mainly lived in feudal France. The best known of them, Quesnay (1694-1774), is specifically interesting in this context, since he used a model of industrial production which is similar to our input-output tables. In his analysis he identified three social classes, namely landowners, peasants, and artisans. His conclusion was that only land created a surplus and therefore only the landowner had the right to obtain it.

Adam Smith (1723-1790) established the structure of economic science. All subsequent thinkers have been guided by concepts as production, value and distribution. These concept will also be of importance in this book.

2.2 Ricardo

Ricardo's main objective (his theoretical objective), which is also the subject matter of Political Economy, was to determine the laws which govern the distribution of income between workers, capitalists, and landowners. These income categories were not arbitrarily selected but corresponded to the three main social classes the way he saw them in England at that time (historical conditioning).[1] These classes can be viewed as sociological data, which Ricardo used when he described the economic system.

The classical economists believed that only land and human labor could be the source of wealth. The organization of production over time required advances since wages had to be paid to the workers before the final product was made (e.g. food). To have capital, therefore, was equivalent to being able to control resources, which could be used to control the workers. From the classical point of view, therefore, capital is a wage fund. (Machines embody previous payments from the wage fund.)

Furthermore, the surplus is the volume of goods which remain when the workers have obtained their share (subsistence minimum).

Below we shall describe a simple Ricardian system (see Pasinetti, 1959-1960) which in spite of its simplicity highlights the problems with which Ricardo worked and which also appear in neo-Ricardian theory.

Assumptions

(i) The economic system only produces one wage good, corn

(ii) It takes exactly one year to produce corn

(iii) Capital consists of the total wage to workers, i.e. circulating capital. (It is not clear how Pasinetti considers seed. It ought to be included in capital.)

---

1) It is not the source of income but a certain relation to the means of production which is decisive for class formation (see Sommerville, 1972, p. 61).

(iv) There exists a numeraire (constant), which we call gold - a luxury good - which at every place and every time requires the same amount of labor for its production (thus constant numeraire). The production of gold takes one year

(v) No fixed capital (machines)

(vi) The land owners' demand rent but they do not produce anything. The capitalists are responsible for production.

The following are given

(a) The amount of land

(b) The quality of land and the technique of farming

We obtain

(1) $X_1 = f(N_1)$ — X_1 = quantity of corn produced during one year; N_1 = number of workers in the corn sector

The production function (1) has the following properties:

(1a) $f(o) \geq 0$ — If no labor is used a negative production is not possible

(1b) $f'(o) > \bar{x}$ — $\bar{x}$ = natural wage rate in terms of corn. Cf. the iron law of wages (which will be dealt with later)

(1b) says that when production is started on the best land, then one must have a surplus; otherwise the economic process will stop

(1c) $f''(N_1) < 0$ — the law of diminishing marginal returns prevail

The production function of gold is simpler:

(2) $X_2 = \alpha N_2$

X_2 = physical quantity of gold produced during one year

N_2 = number of workers in the gold sector

α = physical quantity produced per worker per year

Furthermore we have

(3) $N = N_1 + N_2$ — N = total number of workers

(4) $W = Nx$ — W = total wage in physical units; x = wage in corn

(5) $K = W$ — K = capital stock in physical units

We now come to two important expressions which determine rent and profit which are both expressed in physical units.

(6) $R = f(N_1) - N_1 f'(N_1)$ — R = rent in physical units

(7) $P_1 = X_1 - R - N_1 x$ — P_1 = profit in physical units in the corn sector

(6) and (7) can be illustrated using Figur 2.1 (inspired by Kaldor, 1955-56).

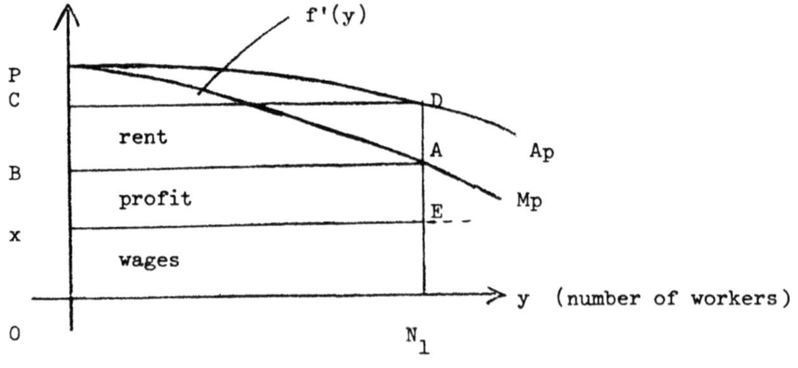

Figur 2.1

The curve P - Ap represents average production per worker. The curve P - Mp represents marginal production per worker. The two different curves are a result of assumption (1c) above.

(6') $$R = f(o) + \int_{o}^{N_1} [f'(y) - f'(N_1)]dy$$

where $f'(N_1) = 0\text{-}B$ and $f'(y)$ are shown in the figure.

By integrating we obtain

(6) $$R = f(N_1) - N_1 f'(N_1)$$

Total production is $f(N_1)$, but it must also be equal to $ON_1 \cdot OC$. The rent, therefore, equals BCDA. Note that marginal land does not receive any rent. (The theory does not need the existence of no rent land since "there were always no rent units of capital and labour at the <u>intensive</u> margin of all land", see Dobb, 1973, p.69.)

The wage per worker is assumed to be given = x, where x in Ricardo's dynamic system will converge towards $\bar{x}$ (see below). Note that x is independent of marginal productivity. Total wage in the corn sector is $N_1 x$ and is equal to $OxEN_1$ in Figure 2.1. Demand for labor is determined by the capital stock K. How many can be employed at the wage rate x?

The profit in the corn sector is determined by

(7) $$P_1 = X_1 - R - N_1 x \quad (= \text{the rectangle } xBAE)$$

The profit, therefore, is a residual when the land owner and worker has received their share. In Figure 2.1 we also can see that when more (and less productive) land is used, i.e. when K is increasing and therefore also N_1 is increasing, then the landowner's share of production will go up while the capitalist's share goes down. This was the result that Ricardo used in his political work (the fight against the corn laws).

Let us return to the original system. Equations (1) - (7) are expressed in physical units. Now we turn to the determination of value.

(8) $$p_1 X_1 - p_1 R = N_1$$ p_1 = the price of corn

(9) $$p_2 X_2 = N_2$$ p_2 = the price of gold

Equations (8) and (9) are important in the Ricardian system. They imply that the value of the yearly production after deduction of rent is determined by the amount of labor necessary to produce it. It is thus a labor theory of value.

The use by the classical economists of the labor theory of value is according to Myrdal (1972, p.41) derived from the natural right doctrine. He finds that the labor theory of value ultimately builds on Hobbes' and Locke's constructions of labor value and property right. Writing about Ricardo, Myrdal (1972, p.105) claims that he never gave any reason for his concept of real value. Ricardo presupposed that his own opinion of what constitutes real value is correct.

Dobb gives a somewhat different interpretation. He is of the opinion that Ricardo constructed his theory of profit before he worked out his theory of value as a foundation of the former (Dobb, 1973, p.32). Later we will return to the problem of value.

From (8) and (9) we can now solve for the prices and obtain

$$(8a) \qquad p_1 = \frac{N_1}{X_1 - R} = \frac{1}{f'(N_1)}$$

$$(9a) \qquad p_2 = \frac{1}{\alpha}$$

The profit in the gold sector, which we have not yet discussed, can be expressed as follows in value terms

$$(10) \qquad p_2P_2 = p_2X_2 - N_2p_1x \qquad\qquad P_2 = \text{profit in physical units}$$

where p_2X_2 is the value of production in the gold sector and N_2p_1x is the cost of production.

Total profit in the whole system is

$$(11) \qquad \Pi = p_1(X_1-R-N_1x) + p_2X_2 - N_2p_1x =$$

$$= p_1X_1 + p_2X_2 - p_1R - p_1W$$

or

$$(11a) \qquad \Pi = (N_1+N_2)(1-xp_1)$$

We now have a theory of value and a distribution theory.

[Rent $= p_1R$
Profit $= (N_1+N_2)(1-xp_1)$
Wage $= p_1W$]

We have, however, not yet connected the flows in the economy. Ricardo assumed that Say's law was true (all income is used).

We only need one equation in our two-sector economy since total production is already determined (cf. (1) and (2)). We obtain

(12) $p_2X_2 = p_1R$ The landowner uses all his income to buy luxury goods, i.e. gold

This is a simplification. The landowners must eat and the capitalists must also eat.

We obtain that $p_1 = p_2 \cdot \frac{X_2}{R}$

i.e. all prices are expressed in gold. The monetary unit is the quantity of gold produced by one worker during one year (cf. (9a)).

Furthermore

(13) $w = p_1x = \frac{x}{f'(N_1)}$ w = the wage in monetary units

(14) $r = \frac{\Pi}{p_1K} = \frac{f'(N_1)}{x} - 1$ r = the rate of profit
Π = the profit share

Ricardo assumed that the rate of profit, r, is the same in all sectors, since if this was not the case, capital would move to the sector with the highest rate of profit. From this follows the fundamental importance of food production in the Ricardian system to determine the level of the rate of profit (cf. (8a)). Ricardo assumed, also when he analyzed more complex systems, that the rate of profit was determined on the margin in the food producing sector (see Dobb, 1973, p.71). When corn production goes up, the rate of profit falls, see (14).

If the price of corn goes up the share of profit falls (see (11a)) and the share of labor goes up

$$\left(\frac{\text{Profit share}}{\text{Labor share}} = \frac{\Pi}{p_1 \cdot W} = \frac{N}{p_1 W} - 1\right)$$

We can now begin to see one of the reasons why Ricardo worked so hard with the value problem. Above it was stated that if the price of corn goes up then the share of profit falls etc, but how is the price of corn measured? (A reason could be that the productivity in the gold sector increased.) When he studied more complex systems, he found that the rate of profit depended upon relative prices and the value of total production was dependent upon the distribution of income between the social classes. This explains Ricardo's search for a constant store of value. Later (in Chapter 3) we shall see how Sraffa helped him at least partly.

If we return to our equations (1) - (14) we see that we have 14 equations but 16 variables. Ricardo closed the system by

(15) $x = \bar{x} > 0$ $\bar{x}$ = subsistence minimum

(16) $K = \bar{K} > 0$ $\bar{K}$ = capital at the beginning of one year

Equations (1) - (16) define an equilibrium; however not a static but a dynamic one. Let us summarize some properties of the system (1) - (16):

(1) It contains a theory of value which is independent of the distribution of income (cf. (8) and (9))

(2) It has a "price theory" which shows that prices depend upon technical factors (see (8a) and (9a))

(3) Wage goods and luxury goods play different roles in the system. The value of all variables except p_2 depends upon $f(N_1)$ or its derivatives,

e.g. $w = \dfrac{\bar{x}}{f'(N_1)}$ $\qquad r = \dfrac{f'(N_1)}{\bar{x}} - 1$

(cf. Sraffa's "basic and non-basic products"; Chapter 3).

(4) An increase of rent reduces the amount of corn which goes to the capitalists. This explains the different points of view of landowners and capitalists. Ricardo was in favor of free trade, this would work against "diminishing return" in farming and be favorable to the capitalists, since corn could be imported (cf. Dobb, 1973, p. 90).

In (15) we assumed that $x = \bar{x}$. We shall study the mechanism which causes this.

Equation (15) is replaced by

(15a) $N = \bar{N}$ $\bar{N}$ = the number of workers available at the beginning of the year

What is given at the beginning of the year is, except $\bar{K}$, available labor and not the wage rate as was assumed in (15).

Ricardo was convinced that the wage rate x, can only temporarily depart from $\bar{x}$ (subsistence minimum). If $x > \bar{x}$ the population will increase and the increase of the labor force will cause wages to go down (the iron law of wages, cf. Malthus). This can be mathematically formulated as follows.

(17) $\frac{dN}{dt} = F(x-\bar{x})$ with the properties

(17a) $F(o) = 0$

$F' > 0$

which means that the population is stable when $x = \bar{x}$. Furthermore it holds that $x = \bar{x}$ is a stable solution only if $\frac{dx}{dN} < 0$ (which is true in the system). Thereby a dynamic mechanism has been introduced into the system). We shall now turn to another dynamic mechanism, namely the accumulation of capital. We begin by studying the steady-state solution - the stationary state - in order to later study the growth process more closely.

The stationary society

Economic growth can, according to Ricardo, be achieved in two different ways, namely

(a) technological development (shifts in the production function), and
(b) accumulation of capital

Compared with his predecessor Smith and his follower Marx, Ricardo puts very little emphasis on technological development. He observes that the effect of capital accumulation (the stationary society) is delayed because of technological development. He considers (b) to be more important than (a), which probably was due to historical conditioning.

Mathematically we can represent the accumulation of capital in the following way

(18) $$\frac{dK}{dt} = \phi\left(\frac{1}{p_1} \cdot \Pi\right);$$ (physical units) or from (8a) and (11a)

(19) $$\frac{dK}{dt} = \phi\left(N\left[f'(N_1) - x\right]\right)$$ with the properties

(20) $\phi(o) = 0$ (stationary when $f'(N_1) = x$,

$\phi'(o) > 0$ i.e. when $\Pi = 0$, cf. Figure 2.1)

The two dynamic mechanisms (17) and (19) cease to operate when $x = \bar{x}$ and $\Pi = 0$.

The stationary society is characterized by

system (1) - (14) plus

(15) $x = \bar{x} > 0$

(16a) $\Pi = 0$

Under certain conditions this solution is unique and completely stable, and the system will converge towards that solution, the stationary society, where the wage is at its natural level $\bar{x}$ and the profit $\Pi = 0$. The condition is

$$(20) \qquad f'(0) > \bar{x} > f'(\infty) \qquad \text{where } f'(\infty) = \lim_{N_1 \to \infty} f'(N_1)$$

(20) says that at a certain point in time when the population is growing, the production of the latest worker must be lower than the wage $\bar{x}$; otherwise the system can expand indefinitely and the stationary society will never be achieved.

The growth process

Above we have observed that the system is stable, but Ricardo also makes an analysis of the growth process itself. The analysis is undertaken under the assumption that the demographic mechanism (17) has been operating but that the accumulation of capital is not complete.

As long as the accumulation of capital goes on, the following increase (see Pasinetti 1959-1960)

(a) number of workers (N)
(b) total wage (W)
(c) total rent (R)
(d) price of corn (p_1)
(e) wage in monetary units (w)

As a result of the same process, r will decrease (r is the rate of profit, which is the same in all sectors).

Ricardo also studied how total profit Π (the share of profit in value units) in the system behaves when the accumulation of capital takes place. One can show that the sign of $\frac{d\Pi}{dK}$ depends upon the size of K

$$\left(\frac{d\Pi}{dK} = \frac{1}{f'(N_1)} \left[\frac{f'(N_1)}{\bar{x}} - 1 + K \cdot \frac{f''(N_1)}{f'(N_1)} \cdot \frac{dN_1}{dK} \right] \right)$$

When K = 0, i.e. at the beginning of the process of accumulation, then $\frac{d\Pi}{dK} > 0$.

In the stationary society where $f'(N_1) = \bar{x}$, we have that $\frac{d\Pi}{dK} < 0$. Inbetween we must have at least one value of K where $\frac{d\Pi}{dK} = 0$, i.e. where the profit takes on its maximum.

Why don't the capitalists stop the accumulation of capital at that value of K? Ricardo introduces a behavioral assumption. It is in the nature of the capitalist to accumulate.

A multi-sector economy with machines (fixed capital)

If the two-sector model with only circulating capital discussed above is extended to a multi-sector model with circulating capital, nothing dramatic will happen; the system is still determined. It is, however, true that if many luxury goods are introduced then total production of luxury goods is determined, but not the amount of individual luxury goods (here individual demand functions are needed; cf. Johansen, 1972). The production of individual luxury goods was of less interest to Ricardo. More serious problems arise, however, when fixed capital is introduced. If we assume uniform composition of capital, then the system is still determined. If this assumption is not fulfilled, the problem is that total employment is not only determined by total K but also by how the demand (and thereby capital) is divided between the different luxury goods.

Ricardo, furthermore, observed that different products require different proportions of material and equipment in relation to work and the machines have different length of life. The competition among the capitalists require that the profit ratio r is equal in all sectors. Total cost for goods which are input in other goods include the profit which the capitalists require on their investment, i.e. the prices depend upon the rate of profit and - what is more difficult - the relative prices depend upon the distribution of income between labor and capital. Assume that the wage rate goes down, whereby the rate of profit goes up. In the labor intensive sectors the share of profit will rise the most. To make the rate of profit the same in all sectors - which is required by competition - the relative prices must change. This is what Dobb (1973, p.84) calls the true nature of Ricardo's problem, namely that it seems as if total production is altered when the distribution of income between the classes changes. Do the prices change because the products have changed (in some sense) or because the numeraire has changed? This explains Ricardo's search for "an invariable standard". Sraffa was going to help him 150 years later (see Chapter 3).

2.3 The reaction against Ricardo

As we saw in the previous chapter, it was a gloomy picture of the world that Ricardo painted. It was a world in which only the landowners could win.

It is easy to understand that Ricardo in his political work fought against the corn laws and thereby against the British landowners. The landowners in their turn fought to hold the cheap foreign corn out of the country. The capitalists could be rather satisfied with Ricardo's theories. They were described as being the cause of economic growth but they could not be held responsible for the low wages of the workers. In Ricardo they found arguments in their fight against the landowners.

The position of the workers was really dark in the Ricardian system but a small beam of light existed, namely the labor theory of value. The Ricardian socialists tried to develop Ricardo's theories further. Hodgskin argued, using the labor theory of value (based on the natural right doctrine), that workers had a right to everything produced (see Dobb, 1973, p. 137).

The critique against Ricardo during his life and after his death was mostly directed against (Marx and the neo-classics will be discussed later)

(a) his theory of value and profit
(b) his theory of rent (some people were afraid of the consequenses of his theory; the position of the landowners was against the interest of society)
(c) his use of Say's law

Critique of the theory of value[1)]

(a) Bailey; Bailey's criticism referred to the concept of absolute value used by Ricardo and thereby his invariable standard. Bailey was interested only in relative values or exchange values. But to define value as relative value or exchange value does not in itself contrast with those who try to find the foundation - or origin - of value and look for this absolute value in the cost of production (Dobb, 1973, p. 102).

(b) The Senior-Longfield group; Senior and Longfield used the utility concept (perhaps due to the influence of Bentham). These two and the group around them tried to develop what now is known as a Cost-of-Production Theory of Value, which one can find in Mill and can trace as far as Marshall. One thereby had a subjective real cost theory (cf. Smith's "tail and trouble"), which could be the foundation of a theory of profit and a theory of wages. Later, however, demand started to play a larger role, whereby the Cost-of-Production Theory of Value became less important. According to Dobb, Ricardo would not have denied that supply and demand has some influence on market prices. However, supply and demand could not be the decisive factor.

Critique of the theory of profit

(a) Longfield; Longfield developed something which resembles a theory of marginal productivity for the profit (the profit depended upon the increased productivity of labor when capital was invested in machines). Profit was the reward for not consuming.

(b) Senior; According to Senior the profit is the reward for abstinence where abstinence represents delayed pleasures. For Senior profit and rent are predetermined and the wage is residual (cf. Ricardo where the profit is residual).

(c) The Senior-Longfield group; As was mentioned above a Cost-of-Production Theory of Value was developed. This subjective

---

1) For a more detailed discussion, see Dobb, 1973, Chapters 4 and 5.

theory of real cost could be the foundation of a theory of profit in addition to a wage theory. The rent remained a residual, since it was hard to imagine any costs associated with the gifts of nature. The important aspect of this new theory was that profit and wages were simultaneously determined. Thereby the field was open for production functions, as well as supply and demand functions (which were to come later). Like the end products, the factors of production are priced by the market mechanism. The sociological data (classes) used by Ricardo to characterize his system disappear into the background. Institutional factors cannot influence the distribution of income between factors of production, but possibly between individuals (see Dobb, 1973, p. 114).

Critique of the use of Say's law

Ricardo used Say's law (all income is spent). This law is built on two assumptions, namely

(a) the demand for goods is infinite
(b) demand was guaranteed, since every good costs something and every cost is somebody's income.

Malthus did not believe in Say's law. He was worried about a surplus of goods. Malthus never succeeded in refuting Say's law but he had sensed the problem of the business cycle. Keynes was later to show the inconsistency of Say's law.

2.4 Marx' system

Marx accepted many ideas behind Ricardo's theory of income distribution. Marx agreed with Ricardo that profit is a residual, but he criticized Ricardo for not carrying the analysis far enough and showed how such a residual can exist and why it goes to the capitalists.

Even if Marx accepted Ricardo's idea that the profit is a residual or a surplus, there were various differences between them. Marx did not use the law of diminishing returns and therefore the difference between rent and profit never became important. Marx regarded the price (supply) of labor as given, not measured in corn but in goods in general.

Both Ricardo's and Marx' wage theories were subsistence minimum theories. However, they gave different reasons for why the wage level was at that minimum. For Ricardo the reason was the iron law. For Marx the reason was the existence of the reserve army of unemployed workers, who prevented the wages rising above the level which was necessary to make the workers do the work. The reserve army was guaranteed by the development of the capitalistic system.

Both Ricardo and Marx assumed that the capitalists accumulated. In Ricardo's system this was a behavioral assumption. In Marx' system, the capitalist had to accumulate because of competition among the capitalists themselves (and economies of scale). If the capitalist did not accumulate, he did not survive.

According to both Ricardo and Marx, the rate of profit falls when capital is accumulated. Ricardo thought that the reason was diminishing return in the farm sector and an identical rate of profit in all other sectors (the latter assumption was taken over by Marx). For Marx the reason was that the ratio between constant and circulating capital grew (this secured the existence of the reserve army).

The results of the dynamic analysis were also different. For Ricardo the stationary society was the result of the accumulation of capital. For Marx the accumulation of capital would lead to the collaps of the capitalistic system. (The reader will find a formal analysis of Marx' main work Capital in Morishima, 1973, where the validity of Marx' results and statements are investigated.)

In the beginning of this section it was mentioned that Marx criticized Ricardo for not having shown how profit is generated. Using the labor theory of value Marx tries to go deeper than Ricardo. How this is done can be summarized in the following points (see Medio, 1972, p. 321).

(a) The economic system can be analyzed at two different levels of abstraction, namely value and price. We can go from the former to the latter in a formal way.

(b) The central concept in the value analysis is abstract work, i.e. work without all individual and specific features, work without considering a special purpose, simply work, average consumption of human energy (Dencik et al., 1969, p. 24). Value is defined as

value = abstract labor time

(c) In the value analysis the commodities are considered as values. If commodities are exchanged according to their values, how is profit, surplus, possible? The solution lies in a commodity which is different from all others. This commodity is labor. As the capitalists exchange commodities according to their values, the workers sell their commodity, labor, at their true value. The true value of the worker is the labor which is necessary to maintain the life of the worker. What makes the capitalist buy labor is its ability to produce a value which is larger than its own value. The difference is surplus value.

Marx defines the degree of exploitation as

$$e = \frac{\text{surplus value}}{\text{the value of labor}} \quad \text{(the same in all industries)}$$

The degree of exploitation which is institutionally and historically determined also serves to close the Marxian system. Furthermore, Marx attempts to show that a positive rate of profit leads to the existence of a positive degree of exploitation and that it is then larger than the rate of profit.

In (a) above we mentioned that one can go from the value system to the price system. This problem, how values are transformed to market prices (the transformation problem) was not completely solved by Marx. Another problem which is connected with the transformation problem is the so called numeraire problem. For Marx it was necessary to find a good or a composite good, which had a value which was equal to its price ("a fixed point problem" in matematical terms). Sraffa was going to lay a foundation for a solution of this problem. (Medio, 1972, has transferred Sraffa's solution to Marx' system.)

We mentioned above that Marx' theory of wages was of a "subsistence minimum" nature. Even if this minimum level does not need to be equal to subsistence minimum in a physical sense, this wage theory was influenced by conditions in England at that time. In a modern society with strong unions, where the wage rate is set by central negotiations, the institutional conditions have changed compared with England as Marx saw it. If one accepts that the structure of the present economic system is different from the one that existed when Marx lived, we can see the Sraffa system (see Chapter 3) as a modification of Marx' system to fit modern society.

2.5 The neo-classical system

Earlier in Section 2.3 we mentioned the early critique on Ricardo. This critique was completed by Jevons et al.

The focus started to shift from the production side to the market and the demand side when one tried to describe the functioning of an economy. The labor theory of value, as it was formulated by Ricardo, and the concept of exploitation belonged to the production side and concerned social relations. The neo-classics concentrated on a theory of relative prices of goods, based on the concept of utility which is a psychological idea. A subjective theory of value was created.

The value theory

Jevons succeeded in solving a problem which the classics had failed to solve, namely the connection between exchange value and use value. By introducing the concept marginal utility he solved the old problem of water and diamonds.

The price of a good was governed by its marginal utility. There was no need to compare total utility or to speak of use value. If there is plenty of water, then its marginal utility is small. The marginal utility of diamonds is high, because diamonds are rare. This, however, led to an unpleasant conclusion. Since marginal utility of every good goes down, as more units are bought, the marginal utility of total income must go down as an individual has more to spend. This means that for a given price of a good, a poor man must have more motivation than a rich man in order to buy it (see Robinson and Eatwell, 1973, p. 36).

Since total utility in society should be maximized (cf. Bentham's philosophy which was the foundation of utility theory), one can see that the neo-classics could have been revolutionary. However, Pareto denied that it was possible to add the utilities of different people. Therefore it could not be proved that total utility would increase if one took one dollar from a rich man and gave it to a poor.

What furthermore happened in the neo-classical break through was that the economic system was to be characterized in a different way and this was due to the theory of marginal utility. The division of society in social classes disappeared, at least for some neo-classicals, and instead individuals with their wishes and needs came to play a decisive role in the analysis. The classical analysis was vertical (to use Robinson's terminology), the neo-classical was horizontal.

Furthermore, the structure of the causal relations were altered drastically. For Ricardo and Marx the distribution of income came before market prices of goods in the sense that prices could only be computed after the determination of the distribution of income. In neo-classical analysis the determination of income distribution was reduced to a pricing problem, where the price of input and output is determined simultaneously in a market process. However, the decisive factor was the demand for final products and thereby the wishes of the consumers.

But, and here appears the dilemma for this type of distribution theory, demand cannot be determined before the resources of the consumers are known. Thereby one needs an assumption about distribution of resources as input to the pricing process, which is supposed to determine the distribution of income.

The distribution of income

The neo-classical distribution theory has its roots in Ricardo's theory of rent. Marx absorbed Ricardo's labor value theory but the neo-classics picked up another element instead, namely the concept of marginal production which Ricardo used to explain the rent.

The difference between Ricardo and the neo-classics was, among other things, that (see Kaldor, 1955-1956)

(a) Ricardo used the principle of substitution (which lies behind all marginal analysis) only with respect to the application of labor on land. In the neo-classical theory it was applied to every factor of production in relation to every other factor. (The principle was also used for goods on the consumption side, why, according to Kaldor, 1955-56, the neo-classical value theory, utility theory, is Ricardo's rent theory applied to consumer demand.)

(b) Ricardo used marginal analysis to show that a fixed factor will get a surplus which is determined by the difference between average production and marginal production, i.e. C-B in Figure 2:1, for the variable factor. The marginal product of labor is equal to wage per capita plus profit per capita, i.e. O-B in Figure 2:1. This was Ricardo's argument.

The neo-classics concentrated on the reverse aspect, i.e. every variable factor receives a reward which corresponds to its marginal product. If we assume that the curve P-Mp in Figure 2:1 denotes the marginal product for an arbitrary variable factor, then this factor will get a price O-B per unit and the rectangle $OBAN_1$ will represent the share of total output received by that factor. The rectangle BCDA represents the reward to all other factors according to their marginal production. This requires that the production function is homogenous of first degree.

One should perhaps also observe that it is not the productivity of a factor but its scarcity relative to other factors which determines the marginal productivity of each of them (see Robinson and Eatwell, 1973, p. 88).

A great difficulty associated with this method appeared when one used capital as a production factor. Land can be measured in hectar per year, labor in man hours, but capital (to be distinguished from capital goods) cannot be measured in physical units. Sraffa finally showed that no profit theory can be based upon the concept "marginal productivity of capital". Capital, the value of the means of production, is not independent of the distribution of income (between labor and capital) as will be shown later.

That part of the production which was given to the workers, the wage, could easily be interpreted in the terms of utility theory. The wage is the reward for the efforts of the worker, a reward which is necessary to accept the "disutility" of labor. We can then obtain a supply curve of labor, see Figure 2:2 (cf. Figure 2:1).

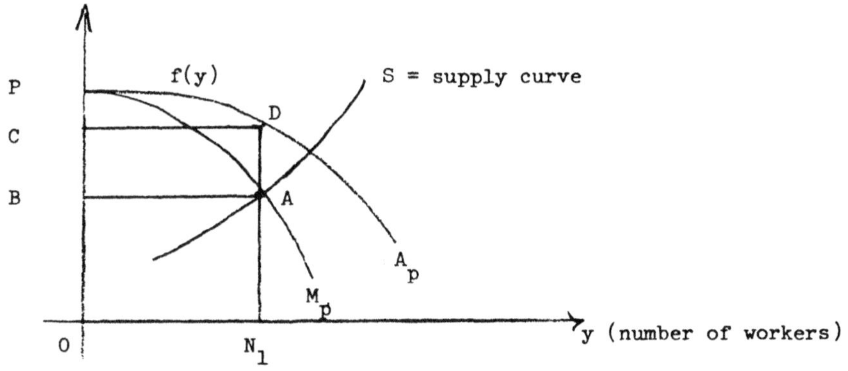

Figure 2.2

The number of workers which will be employed is N, since the marginal product equals the supply price.

But how can the "reward" to other factors of production be interpreted using the concepts of utility theory? The production equipment is productive but it is the owner of the equipment who receives the reward. The share of income going to the capitalist, the return on capital, was the reward because he waited. To wait is to own wealth, the reward goes to the capitalist for his sacrifice of not consuming his wealth (see Robinson and Eatwell, 1973, p.39).

The return on capital was determined by the demand and supply curves. Behind these curves and the neo-classical production function, a serious problem was waiting, namely what was going to be called the possibility of "double switching" and "capital reversing". This destroyed the theory of income distribution based on marginal productivity and the aggregate production function. The connection between capital and productive equipment was broken. The assumption that the value of capital and the profit rate move in different directions proved not to be true. The one who first treated this problem systematically was Sraffa.

Equilibrium analysis

The central theme of the classical economists and Marx was accumulation. As a result of this accumulation they saw different things, Smith a happily growing world, Ricardo the stationary society, and Marx the breakdown of the capitalistic society. The central theme of the neo-classical economists was equilibrium. The interest in finding causal relations diminished. In the Walrasian system everything became dependent on everything else (simultaneous determination). In a modern version (which is described in Klein, 1973, Chapter 9), we meet enterprises which produce products and sell them to the consumers. The interdependence of production (intermediate products) disappears. The enterprises maximize the profit and the consumers maximize utility. The system is closed by the resources of the individuals (in physical units). The resources of the individual are thus exogenously given. The products and the resources are priced in the market process (i.e. wealth is determined). Here we shall not discuss all assumptions which must be fulfilled in order for a price system to exist. But we see the neo-classical dilemma concerning the distribution of income between individuals. One has to assume a distribution of income (resources) in order to be able to determine the distribution of income.

The price system discussed above gives relative prices so we have also here a numeraire problem. Ricardo's old problem reappears. If relative prices change, does this depend upon an alteration of the numeraire?

Furthermore the profit will be zero in long-run equilibrium because of perfect competition. Therefore, a theory of profit is unnecessary. What is left is what de Vylder (1970) calls different points of view regarding the profit.

Demand plays a central role for the specification of relative prices in the neo-classical system. Sraffa was later to show that demand does not influence the rate of profit and relative prices in the long-run equilibrium.

2.6 The Keynesian revolution

Neo-classical theory led to the belief that it was best if the economic system was left to itself. When there was unemployment it was wrong to pay unemployment compensation since it would make the labor unions unwilling to let the wages fall. If only the economic system was left alone, equilibrium and full employment were restored. When the great depression came in 1929 its causes and persistance was an enigma for which the theory provided no solution.

The most fundamental Keynesian critique against both classical and neo-classical theory was the assumption of a timeless equilibrium. Such an equilibrium is possible only if one can foresee the future, at least in the sense that risks can be estimated and that probability theory - in principle - reduces uncertainty to the same level of theoretical analysis as certainty. But if this is not the case and if depressions of the magnitude as the one in 1929 are possible then the logical foundation of equilibrium models are invalidated. A timeless state of equilibrium is impossible.

An important reason why a decrease of the wage rate does not increase the rate of profit and thereby the willingness to invest, and employment was, according to neo-classical theory, the relation between wages and prices.

The uncertainty about the future is the prime reason for the fluctuations in the economic activity. As was mentioned above, the uncertainty is very high and the opinion about the future is of utmost importance. A depression is a situation where profits are low because the willingness to invest is low.

A decrease of the wage rate does not lead to an increase in profits and thus increase in investment etc since a decrease in wages will only reduce prices and mainly leave profits unaffected.

2.7 Conclusions

Ricardo's economic system was the first that was rather complete and it has in various ways inspired various following systems. Marx, for instance, used andextended the labor theory of value, and neo-classical theory has extended and further developed marginal analysis.

There was, however, one major difficulty with Ricardo's system. It was designed for the analysis of income distribution and prices but it had to use something as a "standard" commodity which could be used as a numeraire.

The problem facing Ricardo was that when he wanted to analyze income distribution and prices, he did not know whether the effects he saw were the ones he wanted to study and explain or if what he saw was only a reflection of the fact that the unit of measurement, the standard commodity, had changed.

The very important problem of finding an invariable standard was to be solved by Sraffa.

CHAPTER 3 THE SRAFFA SYSTEM

3.1 Introduction

The main purpose which Sraffa (1960) has in his book is to lay the foundation for a critique of those value- and income distribution theories which build on marginal analysis. (Compare Sraffa's subtitle: Prelude to a critique of economic theory.)

Sraffa's presentation can initially seem somewhat strange, specifically part I which deals with simple production with circulating capital ("Single-Product Industries and Circulating Capital"). One would perhaps have expected a system of equations of the type which are used in input/output analysis. One reason why he departs from the method of description used in input/output analysis is that when he in Part II comes to "joint production" (one production process can give many different products), the latter method of description becomes meaningless.

Another reason for the method of presentation chosen might be that Sraffa wanted to emphasize that he does not assume constant returns to scale. With Sraffa we are in an economy where production continues exactly the same way, year out and year in. (We are in the long run equilibrium.) And without alterations we naturally need no assumptions about constant returns to scale. It should be pointed out, however, that not every economist agree with Sraffa that he does not need to assume constant returns to scale. (For some recent views, see the discussion between Burmeister, Eatwell and Levine in Journal of Economic Literature, March 1977.)

The "alterations" which Sraffa studies, e.g. in the analysis of subsystems and by the construction of the so-called standard system, are no real alterations but only different ways of looking at a given economic system.

Another aspect of Sraffa's system which might seem strange is that he has no demand side. Sraffa shows that in "the long-run equilibrium" prices are only determined by conditions of production. This result is by itself an attack upon neo-classical theory of value, the utility theory. In the neo-classical equilibrium it is necessary, for the existence of a price system, that the consumers maximize their utility.

Furthermore, there are no companies in Sraffa's analysis. Instead we have industries or rather production processes (compare the method of description used by classical economists). The reason for this abstract and simplified method of description of an economic system is that Sraffa wishes to isolate those alterations in the prices which only depend upon alterations in the distribution of income between labor and capital. By this he accomplishes his main purpose, namely to show that the amount of capital (contrary to capital equipment) cannot be measured independently of the distribution of income, which the neo-classics thought was possible (compare the neo-classical production function).

We will mainly deal with section I in Sraffa's book ("Single-Product Industries and Circulating Capital"). The main reason for this is that his most important results appear already there. (The possibility of so-called double-switching, which was first shown by Sraffa and which he deals with in Part III of his book, will be discussed later in Chapter 4.)

By concentrating on Part I, it will also be possible to use matrix algebra which simplifies the understanding of Sraffa's reasoning. (This may seem strange for certain readers, but those who have tried to penetrate Sraffa's work will probably agree.)

From the preface to Sraffa's book it follows that he began his work around 1920. In our presentation we can use results in the theory of matrix algebra which were developed around 1950. (We refer to work by Debreu and Herstein, 1953.)

In section 3.2 we introduce a very simple economic system, characterized by its production without surplus. In section 3.3 a surplus is introduced which goes to capital. In section 3.4 we start a description of the real Sraffa system.

3.2 Production without surplus

We shall study a very simple society which produces three kinds of products - iron, coal, and corn. The products are produced in different industries and are then traded on a market.

To produce corn, totally 240 tons are needed as seed - and to feed the workers. Furthermore - using the current technique - 12 tons of iron and 18 tons of coal are used. All this gives, during a certain time (1 year) 450 tons of corn. In the same way it is possible to write what is needed for the production of iron and coal according to the system of equations for the whole society.

$$\begin{aligned} &240 \text{ tons of corn} + 12 \text{ tons of iron} + 18 \text{ tons of coal} \Rightarrow 450 \text{ tons of corn} \\ &90 \text{ tons of corn} + 6 \text{ tons of iron} + 12 \text{ tons of coal} \Rightarrow 21 \text{ tons of iron} \\ &120 \text{ tons of corn} + 3 \text{ tons of iron} + 30 \text{ tons of coal} \Rightarrow 60 \text{ tons of coal} \end{aligned} \qquad (3.1)$$

If we add the amount of corn, iron, and coal that are used, we see that the sum is equal to the amounts produced.

The different products end up at the end of the production period in the hands of the producers. In order for the production process to continue it is necessary that the commodities are distributed between sectors according to what is shown on the left hand side of equation (3.1). In order for this to take place certain relative prices must exist. If we set the prices such that one ton of iron has the same value as 10 tons of corn, and also the same value as 2 tons of coal, then the different producers will have the opportunity to buy the amounts of raw material necessary to start production again. If we consider the producer of corn, for example, he has 450 - 240 = 210 tons for which he can buy steel and coal. 12 tons of iron cost 120 tons of corn, and 18 tons of coal cost 90 tons of corn, which means that he uses his 210 tons.

What has been described so far can be made more general, and we can prove that it holds for k products and k production processes. Let us use the following notations:

A = total amount produced of product a

A_a = the amount of product a which is used in its own reproduction

B_a = the amount of product b which is used to produce product a

etc.

p_a = the price of one unit of product a

etc.

We obtain the following system of equations for an economic system without surplus:

$$
\begin{cases}
A_a p_a + B_a p_b + \dots\dots\dots\dots + N_a p_n = A p_a \\
A_b p_a + B_b p_b + \dots\dots\dots\dots + N_b p_n = B p_b \\
\cdot \\
\cdot \\
\cdot \\
\cdot \\
A_n p_a + B_n p_b + \dots\dots\dots\dots + N_n p_n = N p_n
\end{cases}
\tag{3.2}
$$

Since the system reproduces itself without surplus we also have that

$$
\begin{cases}
A_a + A_b + \dots\dots\dots\dots A_n = A \\
B_a + B_b + \dots\dots\dots\dots B_n = B \\
\cdot \\
\cdot \\
\cdot \\
\cdot \\
N_a + N_b + \dots\dots\dots\dots N_n = N
\end{cases}
\tag{3.3}
$$

In the system (3.2) we have n unknown (prices, exchange values) but only n-1 <u>independent</u> equations. The system of equations (3.2) has n equations, but if we know n-1 then we also know the n:th due to (3.3).

If we set one exchange value $p_i = 1$, then this is the numeraire of the system, and the exchange values are uniquely determined.

3.3 Production with a surplus which goes to the capitalists

An unrealistic assumption in the formulation used so far is that there is no surplus at the end of the production period. Let us start from the system (3.2) which we can write in the following way:

$$
\begin{cases}
(A_a - A) p_a + B_a p_b + \dots\dots\dots\dots + N_a p_n = 0 \\
A_b p_a + (B_b - B) p_b + \dots\dots\dots\dots + N_b p_n = 0 \\
\cdot \\
\cdot \\
\cdot \\
\cdot \\
A_n p_a + B_n p_b + \dots\dots\dots\dots + (N_n - N) p_n = 0
\end{cases}
\tag{3.4}
$$

If we introduce a surplus in this system (i.e. (3.3) does not hold), then (3.4) is self-contradictory ((3.4) then will be a homogeneous system of equation with the determinant not equal to zero). We have n independent equations with n-1 unknown relative prices.

In order to solve this problem Sraffa introduces the rate of profit r which is the same in all sectors (compare Ricardo and Marx). We can regard the uniform rate of profit as a simplifying assumption or as a result of competition, "long-run equilibrium" (compare Harcourt, 1972, p.181).

As was mentioned, prices cannot be determined in isolation, and therefore Sraffa draws the conclusion that the rate of profit and relative prices must be determined simultaneously.

In order to provide an introduction to the treatment of this, we shall study a society which produces only corn and iron with the following production relations:

$$\begin{array}{l} 280 \text{ tons of corn} + 12 \text{ tons of iron} \longrightarrow 575 \text{ tons of corn} \\ 120 \text{ tons of corn} + 8 \text{ tons of iron} \longrightarrow 20 \text{ tons of iron} \end{array} \qquad (3.5)$$

We here see that all iron is used but that there is a surplus of 175 tons of corn.

We now must find a price which enables the producers to buy each other's goods and to divide the surplus equally between them.

If we consider the production of iron, we find that 20 tons are produced of which 8 must remain for further production of iron. The remaining quantity, 12 tons, can be sold at a price which we set equal to 15 tons of corn per ton of iron. (How this price is computed will be shown below). We now get 180 tons of corn but we only need 120 for the iron production. The surplus is 60 tons, which is 25% of the value, expressed in corn, of the goods going into the production of iron.

The computations are done as follows. Let the value of one ton of corn be equal to one, $p_a = 1$. We then get the following system of equations:

$$\begin{cases} (280 + 12\, p_b)(1 + r) = 575 \\ (120 + 8\, p_b)(1 + r) = 20\, p_b \end{cases} \qquad (3.6)$$

This is a non-linear system of equations which has the solution

$r = 0.25$

$p_b = 15$

($p_a = 1$ by assumption)

The general formulation for a society with n products and n production processes is

$$\begin{cases} (A_a p_a + B_a p_b + \dots\dots\dots + N_a p_n)(1+r) = Ap_a \\ (A_b p_a + B_b p_b + \dots\dots\dots + N_b p_n)(1+r) = Bp_b \\ \cdot \\ \cdot \\ \cdot \\ \cdot \\ (A_n p_a + B_n p_b + \dots\dots\dots + N_n p_n)(1+r) = Np_n \end{cases} \tag{3.7}$$

If we assume that the workers do not receive any share of the surplus, i.e. labor is regarded as equal to material methods of production, then (3.7) resembles Marx's system. The wage is as advance payment, the surplus is analogous to Marx's surplus value.

The existence of a surplus makes it possible to produce goods which are not directly required to produce other goods, so-called luxury goods (compare the section about Ricardo, where the landowner bought gold for his surplus, the rent). These commodities play a completely passive role and do not influence the price formation in the economic system. The reason is easy to see. Consider the system of equations below which is system (3.7) extended by a number of equations referring to luxury goods,

$$\begin{cases} (A_a p_a + B_a p_b + \dots\dots\dots\dots + N_a p_n)(1+r) = Ap_a \\ (A_n p_a + B_n p_b + \dots\dots\dots\dots + N_n p_n)(1+r) = Np_n \\ (A_s p_a + B_s p_b + \dots\dots\dots\dots + N_s p_n + S_s p_s + \dots .+V_s p_v)(1+r) = Sp_s \\ \vdots \\ (A_v p_a + B_v p_b + \dots\dots\dots\dots + N_v p_n + S_v p_s + \dots .+V_v p_v)(1+r) = Vp_v \end{cases} \tag{3.8}$$

where products s,....v are luxury goods which are not inputs in the n first equations, where the n first equations constitute the system (3.7).

From (3.8) we see that the prices $p_a,\ldots p_n$ are independent of the existence of luxury goods in the system. However, the prices $p_s,\ldots p_v$ will be influenced by the prices $p_a,\ldots p_n$.

We have now arrived at Sraffa's distinction between basic products and non-basic products. Basic products are those which are used in the production of all other products (directly or indirectly). Other goods are non-basic. As we have seen above basic products will have a decisive role in the determination of prices in the Sraffa system (compare Ricardo in Chapter 2, where the farm sector determined the prices).

3.4 Production with a surplus: the Sraffa system

In the previous sections of this chapter, labor has not been analyzed explicitly. When labor is introduced, certain problems arise. That part of the wage which is necessary for the reproduction of the work force is basic in the same way as coal and oil are basic for the production of other products, while the consumption which goes beyond that constitutes non-basic goods.

Sraffa here chooses to regard the wage as non-basic, and this is essential. The result is that the whole wage is regarded as variable. Furthermore, the wage is no longer regarded as an advance but is paid at the end of the period of production. The products which the workers need to survive are no longer input in the production process but they are replaced by homogeneous labor.

Let us introduce the following notations in addition to the ones already introduced.

L_a = the amount of labor which is necessary to produce the product a

w = wage

r = the profit (expressed as a fraction of the goods used in the production).

We can now formulate the following relations:

$$\begin{aligned}
&(A_a p_a + B_a p_b + C_a p_c \;\ldots\; + N_a p_n)(1+r) + wL_a = Ap_a \\
&(A_b p_a + B_b p_b + C_b p_c \;\ldots\; + N_b p_n)(1+r) + wL_b = Bp_b \\
&\quad\vdots \\
&(A_n p_a + B_n p_n + C_n p_n \;\ldots\; + N_n p_n)(1+r) + wL_n = Np_n
\end{aligned} \tag{3.9}$$

We then normalize labor as follows:

$$L_a + L_b + \ldots L_n = 1 \tag{3.10}$$

We assume that the system can renew itself and thus that $A_a + A_b + \ldots + A_n \leq A$ for all products a,b,...,n. What remains when the amount of the different products used in next year's production has been deducted, is called national income. We use this as a unit of value to express the different prices and wages. We thus obtain the following equation

$$\begin{aligned}
&[A - (A_a + A_b + A_c + \ldots + A_n)]p_a + \\
&+ [B - (B_a + B_b + B_c + \ldots + B_n)]p_b + \\
&+ \ldots [N - (N_a + N_b + N_c + \ldots + N_n)]p_n = 1
\end{aligned} \tag{3.11}$$

An advantage in regarding the whole wage as variable is, according to Dobb, that the net income of the system, i.e. (3.11), will be related to the usual definition of national income, which includes both wages and profit (see Dobb, 1973, p.262).

The national income (3.11) thus constitutes the surplus which shall be distributed between capital and labor, or between capitalists and workers. Dobb (1973, p. 262) regards this assumption, i.e. that workers and capitalists share a surplus, as a possibility to consider the influence of unions in wage negotiations.

Equations (3.9) and (3.11) give us n+1 equations and n+2 variables (n prices, the rate of profit r, and wage w). We thus have one degree of freedom and as soon as the value of one variable is determined the value of the others is determined. If $w = 1$, the whole national product goes to wages, which means that the profit = 0, i.e. $r = 0$. If $w = 0$ then r is maximal (in this latter case we are back to system (3.7), disregarding that (3.7) includes consumption goods).

We can thus close the Sraffa-system by determining what part of national income shall go to labor. In a later chapter we shall see how we can use a result obtained by Kaldor to close the Sraffa-system, i.e. to determine the value of r.

For a given value of r (or w) the prices of goods can be determined in three different ways, namely

(a) by solving the system of equations (3.9) - (3.11)
(b) by using subsystems
(c) by transforming into labor terms

We shall discuss (a) and (b) using an example which is constructed by Bose (1964).

We assume an economy which produces two products, A and B, using these products and labor L. This economy can be described in the following way.

$$\begin{cases} 2A + 2B + \frac{3}{4} L \rightarrow 8A \\ 2A + 5B + \frac{1}{4} L \rightarrow 8B \end{cases} \qquad (3.12)$$

Gross production in (3.12) is 8 units of A and 8 units of B. Net production is

$$\begin{cases} 8 - (2+2) = 4 \text{ units of A} \\ 8 - (2+5) = 1 \text{ unit of B} \end{cases}$$

We can write (3.12) analogous to (3.9). We get

$$\begin{cases} (2p_a + 2p_b)(1+r) + \frac{3}{4} w = 8p_a \\ (2p_a + 5p_b)(1+r) + \frac{1}{4} w = 8p_b \end{cases} \qquad (3.13)$$

The amount of labor is already normalized. As the numeraire of the system we use net national product (national income), compare (3.11).

$$4p_a + p_b = 1 \qquad (3.14)$$

We have three equations and four independent variables.

$$\begin{cases} r = 0 \\ p_a = \frac{11}{56} \\ p_b = \frac{3}{14} \\ w = 1 \end{cases} \qquad (3.15)$$

The value of gross production is

$$8 \cdot \frac{11}{56} + 8 \cdot \frac{3}{14} = \frac{23}{7} = 3\,\frac{2}{7}$$

If we assume the other extreme and give all to capital we obtain

$$\begin{cases} w = 0 \\ p_a = \frac{1}{6} \\ p_b = \frac{1}{3} \\ r = \frac{1}{3} \end{cases} \qquad (3.16)$$

Using (3.15) the value of gross national product is

$$8 \cdot \frac{1}{6} + 8 \cdot \frac{2}{6} = 4$$

The solutions (3.15) and (3.16) to the system (3.13) - (3.14) give us an example of how relative prices are changed when income distribution is altered. We also find a justification for Ricardo's problem that the value of the total national product is altered when the distribution of income is altered. How Sraffa solved this problem shall be discussed in the next section.

We end this section by solving (3.13) - (3.14) using subsystems. This means that system (3.12) is divided into as many minieconomies as there are products in the surplus. In the example used we will obtain two minieconomies where every economy reproduces itself and has only one product in its surplus. Such minieconomies Sraffa called subsystems (see Sraffa, 1960, Appendix A). (For a more complete discussion of subsystems, see Harcourt and Massaro, 1964.)

Let us start with the case $r = 0$, i.e. $w = 1$. We will then get the following system of equations

$$\begin{aligned} 2p_a + 2p_b + \frac{3}{4} &= 8p_a \\ 2p_a + 5p_b + \frac{1}{4} &= 8p_b \end{aligned} \qquad (3.17)$$

The problem is to find weights x and x' (where $x' = 1 - x$) to be used in equation 1 in (3.17) and weights y and y' (where $y' = 1 - y$) to be used in equation 2 in (3.17) which will give us two subsystems.

The weights

$$x = \frac{1}{7} \qquad y = \frac{3}{7} \quad \text{give us one subsystem (system I)}$$

$$x' = \frac{6}{7} \qquad y' = \frac{4}{7} \quad \text{give us one subsystem (system II)}$$

Multiply the first equation in (3.17) with $x = \frac{1}{7}$ and the other equation with $y = \frac{3}{7}$. This will give

$$\frac{2}{7}p_a + \frac{2}{7}p_b + \frac{3}{28} = \frac{8}{7}p_a$$

$$\frac{6}{7}p_a + \frac{15}{7}p_b + \frac{3}{28} = \frac{24}{7}p_b$$

If these two equations are added, we obtain

$$\frac{8}{7}p_a + \frac{17}{7}p_b + \frac{6}{28} = \frac{8}{7}p_a + \frac{24}{7}p_b$$

This expression can be simplified to

$$\frac{3}{14} = p_b$$

This means that in order to produce one unit of b, $\frac{3}{14}$ units of time are required (the third term in the expression above has the dimension time since w = 1). In an analogous way we obtain system II by using the weights x' and y'. We obtain

$$\frac{12}{7}p_a + \frac{12}{7}p_b + \frac{18}{28} = \frac{48}{7}p_a$$

$$\frac{8}{7}p_a + \frac{20}{7}p_b + \frac{4}{28} = \frac{32}{7}p_b$$

$$\frac{20}{7}p_a + \frac{32}{7}p_b + \frac{22}{28} = \frac{48}{7}p_a + \frac{32}{7}p_b$$

$$\frac{11}{56} = p_a$$

Using subsystems we have by an example shown an important result obtained by Sraffa, namely if w = 1 we get a labor theory of value, i.e. we can determine the time (directly and indirectly) which is needed to manufacture one unit of a product.

3.5 The Standard Product

In order to simplify the presentation we shall divide each equation in (3.9) by the quantity on the right hand side, i.e. we divide the first equation by A, the second by B etc. Thereafter we make the corresponding alterations of equations (3.10) and (3.11).

This will give us an economic system where the amount of output of every product is one. Furthermore we have that

$$\sum_i a_{ij} \leq 1 \quad \text{for } j = 1,2,\ldots,N$$

a_{ij} is thus an element in the input-output matrix which we have created by the divisions mentioned above. a_{ij} is the amount of product j which is necessary to produce one unit of product i. (It should perhaps be mentioned that the divisions above do not imply an assumption about constant returns to scale. We only change the units of measurement.)

The profit in the i:th industry r_i can be written

$$r_i = \frac{p_i - \sum_j a_{ij} p_j}{\sum_j a_{ij} p_j} \tag{3.18}$$

As we mentioned earlier it is reasonable to assume that the profit is identical in all industries. We can therefore sum (3.18) over i and write $r_i = r$ $(i = 1.2....,N)$

$$r = \frac{\sum_i p_i - \sum_i \sum_j a_{ij} p_j}{\sum_i \sum_j a_{ij} p_j} = \frac{\sum_i p_i - v}{v} \tag{3.19}$$

where v is the total value of all "raw" material.

If we are now interested in investigating how alterations in the distribution of income influence prices, the problem facing Ricardo reappears. Assume that wages go up. Those industries which use much labor will now be in a worse position than before, compared with those industries which employ few workers. Since the profit must be the same in all industries and since no alteration of the technique is assumed to take place, the prices must alter. An analysis of this is made more complicated by the fact that every industry contains products from other industries as raw material. The fact that an industry has many workers need not be a disadvantage if it is using raw material from industries which utilize little labor. Furthermore, we need a unit of value which does not alter when the prices change. Since if the unit of value can change, we do not know if the alteration in the price of a certain product depends on the fact that its relative value has changed or if it is the unit of value that has changed.

It is here that Sraffa makes an important contribution by showing how one can construct a so called standard commodity by using the existing products in such a way that the value of the standard commodity does not change when the ratio of profit and wages changes.

We look for a product which has the property that the ratio between all what is produced of it and what is used of it in the production process is equal for all products. This means that we look for multipliers q_i ($i = 1,2,\ldots,N$) such that

$$\frac{q_1}{\sum_i q_i a_{i1}} = \frac{q_2}{\sum_i q_i a_{i2}} = \ldots \frac{q_N}{\sum_i q_i a_{in}} = \frac{1}{V} \tag{3.20}$$

The ratio in (3.20) are homogenous in the sense that for example the first ratio has physical units of product 1 (output of product 1 in the standard system) in the numerator which is the same as the denominator (total amount of good 1 which is used as input in the standard system). The ratios in (3.20) are "pure" in the sense that no values are involved. (Another "pure" ratio is direct labor divided by indirect labor which Sraffa did not use.)

(3.20) can be written as follows

$$\sum_i q_i a_{ij} = V q_j \qquad (j = 1,2,\ldots,N) \tag{3.21}$$

Sraffa then defined the standard system as the system which uses the same quantity of labor as the real system. This means that we have one more equation

$$q_1 \cdot \ell_1 + q_2 \ell_2 + \ldots + q_N \ell_N = 1 \tag{3.22}$$

We see directly that the ratio (3.20) is independent of existing prices and therefore they are independent of price alterations, e.g. between labor and capital. This can directly be seen in (3.20) where - if prices are introduced - the expression gets the following form

$$\frac{p_1 q_1}{p_1 \sum_i q_i q_{i1}} = \frac{p_2 q_2}{p_2 \sum_i q_i a_{i2}} = \frac{1}{V}$$

The system which we have now constructed produces a net income which we call the standard net income. As was shown in (3.20) we have constructed the system such that the surplus is the same in all industries.

We shall call this surplus R.

$$R = \frac{\sum_j p_j(q_j - \sum_i q_i a_{ij})}{\sum_j p_j \sum_i q_i a_{ij}} \qquad (3.23)$$

If we use (3.21) this can be written

$$R = \frac{(1 - V)}{V} \cdot \frac{\sum_j p_j q_j}{\sum_j p_j q_j} = \frac{1 - V}{V} \qquad (3.24)$$

If we compare the computations of the weights q in (3.20) with the formulation of the problem in (3.9), we see directly that R corresponds to r in the real system when the whole surplus goes to capital[1], e.g. when w = 0.

When constructing the standard commodity we have multiplied with constants q and we must emphasize that in spite of this we have not assumed that returns to scale are constant. This is the case since the standard commodity is only an artificial construction without any correspondence in the real world.

Assume now that a certain fraction α of the standard income goes to labor. This means that the numerator in (3.23) shall be multiplied by $(1 - \alpha)$ and consequently the share of national income going to capital is determined by

$$r = R(1-\alpha) \qquad (3.25)$$

We shall now show that the expression (3.25) also holds for the real system when the wages are paid in the standard commodity. Call the rate of profit in the real system r_a.

Using a compact expression we can write (3.21)

$$q'A = Vq' \qquad (3.26)$$

---

1) From (3.9) we have for w = 0 that $Ap = \frac{1}{1+r}$. Equation (3.21) can be written $A'q = Vq$. The dominant roots of A and A' are the same and thus $V = \frac{1}{1+r}$. For further discussions of this, see Newman (1962).

where

$$q = \begin{pmatrix} q_1 \\ q_2 \\ \cdot \\ \cdot \end{pmatrix}$$ and q' is the transposed q-vector.

A is a matrix with elements a_{ij}.

Transposing (3.26) gives

$$A'q = Vq \tag{3.26a}$$

From (3.24) we know that $R = \frac{1 - V}{V}$, which gives

$$(1+R)A'q = q \tag{3.27}$$

(3.27) is together with (cf. (3.22))

$$L'q = 1 \tag{3.28}$$

the standard system which is embedded in the real system

$$Ap(1+r_a) + \alpha L = p \tag{3.29}$$

where

$$L = \begin{pmatrix} \ell_1 \\ \ell_2 \\ \cdot \\ \cdot \end{pmatrix} \quad \text{and } p = \begin{pmatrix} p_1 \\ p_2 \\ \cdot \\ \cdot \end{pmatrix}$$

Furthermore we can write

$$z = R(q-z) \tag{3.30}$$

where z is a vector which gives the net output and where

$$z'p = 1 \tag{3.31}$$

i.e. the value of the net national product is 1.

Premultiplication of (3.29) by q' gives

$$q'Ap(1+r_a) + \alpha q'L = q'p \tag{3.32}$$

Using (3.28) we get

$$q'Ap(1+r_a) + \alpha \cdot 1 = q'p \tag{3.33}$$

Multiplying and dividing by 1 + R gives

$$\frac{q'Ap(1+r_a)(1+R)}{1+R} + \alpha = q'p \tag{3.34}$$

By transposing (3.27) we obtain

$$(1+R)q'A = q' \tag{3.35}$$

If (3.35) is inserted in (3.34) we obtain

$$\frac{q'p(1+r_a)}{1+R} + \alpha = q'p$$

$$\alpha = (1 - \frac{1+r_a}{1+R})q'p \tag{3.36}$$

From (3.30) we get

$$q' = \frac{1+R}{R} z'$$

which inserted in (3.36) gives

$$\alpha = (1 - \frac{1+r_a}{1+R}) \frac{1+R}{R} z'p \tag{3.37}$$

where $z'p = 1$ (see 3.31)

(3.37) therefore can be simplified as

$$r_a = R(1-\alpha) \tag{3.38}$$

Observe that α is the share of the standard income that goes to the worker and <u>not</u> the w which refers to the real system (3.9). But <u>if</u> we measure the wage as a share of the standard national product, i.e. $w = \alpha$, then we obtain

$$r = R(1-w) \tag{3.39}$$

where $r = r_a$.

Now we also see that we need not know the standard product (its physical components) to compare the prices using the standard income as numeraire. From (3.39) we see that also the rate of profit depends upon technical factors (R denotes in some sense the productivity of the system) as well as social factors (i.e. the distribution of income between labor and capital). One should note that Sraffa would not have obtained the simple expression (3.39) if he, like Marx, had assumed that wages were paid in advance. (For a translation of the Sraffa system in Marxian terms, see Medio, 1972.)

We can rewrite (3.39) as follows

$$\frac{1}{w} = \frac{R}{R-r} \tag{3.40}$$

In (3.40) we have Sraffa's value theory. The value is the amount of labor that we can buy for the standard national income and as we see from the right hand side in (3.40) the value will vary with r. This means that

> "all the properties of 'an invariable standard of value' ... are found in a variable quantity of labour, which, however, varies according to a simple rule which is independent of prices." (Sraffa, 1960, p.32)

3.6 The Sraffa system: uniqueness

In this section we shall discuss

(a) if R is uniquely determined

(b) if the price vector p is non-negative

Let us begin by a theorem from matrix algebra (see for example Klein, 1973, p.350, Theorem 19.1).

Theorem 1

Let A be a n x n non-negative indecomposable matrix. Then we have that

(1) A has one characteristic value (eigenvalue) $v^* > 0$

(2) If q^* is a characteristic vector which corresponds to v^* then $q^* \gg 0$

(3) For every other characteristic value v for A we have that $v^* \geq |v|$

(4) The value of v^* increases with the increase of an arbitrary element a_{ij} in A

(Fröbenius' theorem)

Here we are interested in an economic system which consists only of basic products. Then the matrix A in (3.29) is indecomposable. (For a formal definition of indecomposable matrices see Klein, 1973, p.346, Def. 19.1.) Our matrix A from last section thus fulfils the requirements of the theorem.

We need one more theorem (see Klein, 1973, p.354, Theorem 19.6).

Theorem 2

Let A be a non-negative indecomposable matrix. Then the matrix

$$(\mu I - A)^{-1}$$

is positive if and only if $\mu > v^*$

(where B^{-1} indicates the inverse of B)

Let us now return to our economic system. In compact form it can be written (cf. (3.29))

$$Ap(1+r) + wL = p \tag{3.41}$$

This in turn can be written as

$$wL = [I - A(1+r)]p$$

$$p = \left[\frac{1}{1+r} I - A\right]^{-1} \cdot \frac{w}{1+r} L \tag{3.42}$$

Since $\frac{w}{1+r} L$ is non-negative we see, using Theorem 2, that the price vector p is non-negative if and only if

$$\frac{1}{1+r} > v^* \tag{3.43}$$

Furthermore, since $r \geq 0$ and we look for v^*, the largest of all possible v, this means that we look for the smallest of all R (i.e. r_{max}). Only this value of R gives non-negative prices for

$$0 \leq r < R \tag{3.44}$$

We conclude this section with an example where we use the numbers in (3.13).

Using (3.13), matrix A can be written

$$A = \begin{pmatrix} \frac{2}{8} & \frac{2}{8} \\ \frac{2}{8} & \frac{5}{8} \end{pmatrix} = \begin{pmatrix} \frac{1}{4} & \frac{1}{4} \\ \frac{1}{4} & \frac{5}{8} \end{pmatrix}$$

The eigenvalue v associated with A is determined as (see Klein, 1973, Chapter 16)

$$\begin{vmatrix} \frac{1}{4} - v & \frac{1}{4} \\ \frac{1}{4} & \frac{5}{8} - v \end{vmatrix} = 0$$

We get

$$v^2 - \frac{7}{8} v + \frac{3}{32} = 0 \tag{3.45}$$

(3.45) has two solutions

$$v_1 = \frac{3}{4} \qquad v_2 = \frac{1}{8}$$

We know that v^* is the largest v-value, i.e. $v^* = \frac{3}{4}$. From (3.43) we obtain $R = \frac{1}{3}$.

Using (3.27) and (3.28) we can determine the q-vector which gives us the standard system.

We obtain $q_1 = \frac{4}{5}$ and $q_2 = \frac{8}{5}$.

The standard system then is (equation 1 in (3.12) is multiplied by q_1, equation 2 with q_2)

$$\frac{8}{5}A + \frac{8}{5}B + \frac{3}{5}L \rightarrow \frac{32}{5}A$$

$$\frac{16}{5}A + 8B + \frac{2}{5}L \rightarrow \frac{64}{5}B$$

$$\frac{24}{5}A + \frac{48}{5}B + 1L \rightarrow \frac{32}{5}A + \frac{64}{5}B$$

The standard system has a gross production of

$$\frac{32}{5} \text{ of A} \quad \text{and} \quad \frac{64}{5} \text{ of B} \tag{3.46}$$

The net production, i.e. the physical components of the standard product, consists of

$$\frac{8}{5} \text{ of A} \quad \text{and} \quad \frac{16}{5} \text{ of B} \tag{3.47}$$

If we solve (3.13) and use the relation $r = R(1-w)$ which gives us the standard national income as numeraire, we obtain

$$\begin{cases} w = 1 \\ p_a = \frac{11}{56} \\ p_b = \frac{3}{14} \\ r = 0 \end{cases} \tag{3.48}$$

for

$$\begin{cases} w = \frac{2}{3} \\ p_a = \frac{43}{248} \\ p_b = \frac{14}{62} \\ r = \frac{1}{9} \end{cases} \tag{3.49}$$

and for

$$\begin{cases} w = 0 \\ p_a = \frac{1}{8} \\ p_b = \frac{1}{4} \\ r = \frac{1}{3} \end{cases} \tag{3.50}$$

We see how the relative prices change when the distribution of income is altered. This happens in spite of the fact that the prices are expressed in the standard national income which is independent of income distribution.

The value of the gross production (3.46) in the standard system is 4, independent of what prices are used in (3.48) - (3.50). This result can be compared with the values of gross production which we obtained in Section 3.4.

3.7 Transformation into labor terms

In Section 3.4 it was pointed out that Sraffa suggests three different ways to compute prices

(a) by solving systems of equations in a traditional way

(b) by using subsystems

(c) by transforming into labor terms

In this section we shall discuss the third method.

As we have seen earlier, we can write the production of a product A in the following way

$$(A_a p_a + B_a p_b + \dots + N_a p_N)(1+r) + L_a w = Ap_a$$

Every product in the left hand side parenthesis can be replaced by an equation which describes its own production.

If we replace the first term in the parenthesis we obtain

$$\frac{A_a}{A}\left[(A_a p_a + B_a p_b + \ldots + N_a p_N)(1+r) + L_a w \cdot \frac{A_a}{A} + \right.$$
$$\left. + B_a p_b + \ldots + N_a p_N\right](1+r) + L_a w = Ap_a \qquad (3.51)$$

As we see this can go on as long as we like.

For every product we receive a labor term multiplied by (1+r), then by $(1+r)^2$, etc. This has the effect that we eventually can write the equation as follows

$$L_a w + L_{a_1} w(1+r) + \ldots + L_{a_n} w(1+r)^n = Ap_a \qquad (3.52)$$

We know that if the wage is expressed in the standard commodity, then the share of wages w is determined by

$$w = 1 - \frac{r}{R} \qquad (3.53)$$

If we insert (3.53) in the different terms in (3.52) these will receive the form

$$L_{a_n}(1 - \frac{r}{R})(1+r)^n \qquad (3.54)$$

The important result of this breaking down of the process in labor terms is that it is the basis for the critique of the neo-classical theory of capital to which we shall devote the next chapter.

Assume that we have two products which differ with respect to the amount of labor that they use at three different points of time.

Call one of the products a and the other b. Product a used 20 labor units more than product b 8 years ago, while b used 19 labor units more this year and one labor unit more 25 years ago. (This example is used by Sraffa and might refer to wine and oak chests.)

The difference between the standard prices is

$$p_a - p_b = 20w(1+r)^8 - \{19w + w(1+r)^{25}\}$$

If we use $w = 1 - \frac{r}{R}$ and put $R = 0.25$ we obtain the curve shown in Figure 3.1.

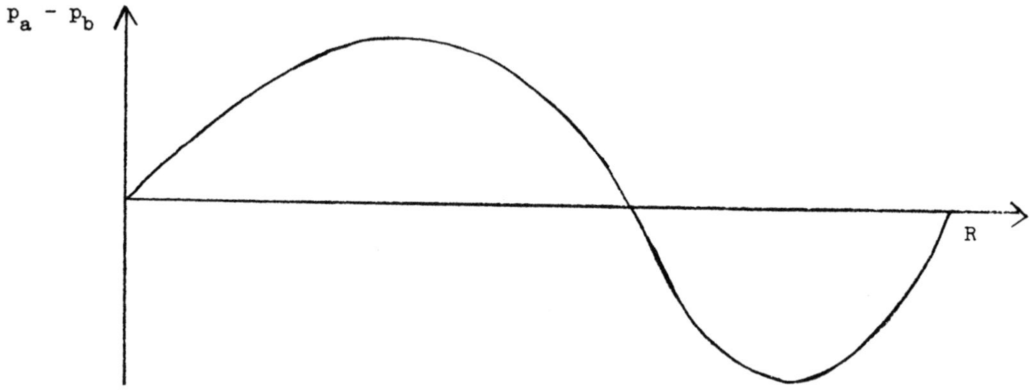

Figure 3.1

The price of wine, a, first goes up when r goes up, then goes down after r = 9 %, and is - when r = 17 % - lower than the price of oak chests (see Figure 3.1). After r = 0.22 it rises again. In the following chapter we shall show what fundamental importance this has for capital theory.

3.8 Conclusion

An assumption in the previous section was that every production process produced only one product and used only circulating capital (no fixed capital).

Part II of Sraffa's book is devoted to joint production, i.e. when a given production process can produce many different products. The case with fixed capital can then be used as a special case of joint production.

We shall not treat these cases in detail, since they do not influence Sraffa's main contribution, namely the critique of neo-classical theory of capital (see, however, Meek, 1974, who thinks that it is for the theory of value that Sraffa will have the greatest influence).

There are, however, some reason to indicate certain problems in the case of joint production. The definition of basic products is not as simple as before. It is possible to get negative prices. (This is analogous to the case with joint production studied by Marx which can give negative labor values, see Morishima, 1973, p.182.) Furthermore, prices cannot be computed by transformation into labor terms.

We conclude this chapter by summarizing Sraffa's main assumptions and results.

The assumptions in Sraffa's system are

(1) The economy being studied produces enough to maintain the existing level of production (self-replacing state).

(2) We study one single point on the productions function. No assumption about constant returns to scale is made and in this respect Sraffa differs from Walras-Cassel-Leontieff whose formulations resemble the one constructed by Sraffa.

(3) The rate of profit is the same in all industries.

(4) The wage is the same in all industries.

(5) When the distribution of income is altered, it is assumed that output does not alter.

Sraffa's book leads up to three fundamental results, namely

(1) If there exist technical coefficients for the production of single products and if the wage rate is equal in the economy, then there is no place for demand functions in price determination.

(2) Prices do not break down in a natural way in profits, wages, and rent. There is a maximum value of the profit but otherwise the system is undetermined.

(3) The concept "capital" cannot be used to explain the distribution of income.

CHAPTER 4 THE CAPITAL CONTROVERSY[1)]

It has been considered obvious and natural that when the time necessary to produce a product goes up, then the number of workers necessary for the production goes down and therefore the production per capita goes up and the rate of interest goes down.[2)] As was pointed out in Chapter 3, Sraffa has shown that it is not clear that a lowering of the interest rate will lead to an increase in the time necessary for production.

To summarize the idea behind the traditional view one can begin by considering two goods; one produced with much labor which has been applied recently and another using less labor which has been applied longer ago.

If the wage rate is low and the interest rate is high, one would expect that the former is cheaper. As the wage rate goes up and the interest goes down it seems natural that the latter will be cheaper since labor costs are cheaper for that good. When the rate of interest goes down capital deepening takes place. This is the traditional viewpoint and in this case one can use a conventional production function.

Assume, however, as was shown in Chapter 3, that we have a situation where one commodity has its main labor input at a point of time t_1 while another has its labor input at t_2 and t_3 and that $t_2 < t_1 < t_3$. It is now possible that the former commodity is better both at high and low rates of interest.

4.1 Determination of the factor-price curve

We shall here study a simple case with only two products and begin with the example discussed in Chapter 3, where we can write the conditions governing production as follows

---

1) The reader will find an extensive treatment of this topic in Harcourt (1972). A recent discussion can be found in Bliss (1975).

2) A more roundabout technique will be better, the lower the rate of interest is.

$$(A_b + B_b p_b)(1 + r) + L_b w = Bp_b \tag{4.1}$$

We have here taken commodity A as a standard of value. From the system of equations above we can solve for w as a function of r and then obtain the following expression

$$w = \frac{(A_a B_b - B_a A_b)(1+r)^2 - (BA_a + AB_b)(1+r) + AB}{(L_b B_a - L_a B_b)(1+r) + L_a B} \tag{4.2}$$

From equation (4.1) it follows that the relation between the wage rate, w, and the rate of profit, r, is as a rule not a straight line. As will be shown below it is, however, of special interest to investigate the case when the relation is linear. A linear relation between w and r is obtained if r is set equal to zero in (4.1) and p_b and w are solved for. It can then be shown that the price is proportional to the labor input in the commodities, i.e. a labor theory of value holds. Furthermore, we have that the ratio of the value of labor and the value of capital is the same in the two production processes, which means that they have the same "organic composition" of capital in the two industries, which is important in Marxian economics.

Example:

To illustrate this further we shall use the following example from Robinson and Naqvi (1967) where t stands for ton.

$$1 \text{ t corn} + \frac{14}{5} \text{ t copper} + \frac{3}{5} \text{ labor} \longrightarrow 3 \text{ t corn}$$

$$1 \text{ t corn} + \frac{6}{5} \text{ t copper} + \frac{2}{5} \text{ labor} \longrightarrow 4 \text{ t copper}$$

which we write

$$(1 + \frac{14}{5} p_k)(1 + r) + \frac{3}{5} w = 3 \tag{4.3}$$

$$(1 + \frac{6}{5} p_k)(1 + r) + \frac{2}{5} w = 4p_k \tag{4.4}$$

where p_k is the price of copper using corn as a numeraire.

Observe that all copper is used and that 1 ton corn remains to be divided between labor and capital. When $r = 0$ 1 ton corn is allocated to labor and $w = 1$ (the unit being corn), since total amount of labor is 1.

We also have to determine the price of copper which we can do from the first equation, for example.

$$1 + \frac{14}{5} p_k + \frac{3}{5} = 3$$

$$p_k = \frac{1}{2}$$

We can now solve for the r - w relation from, for example, (4.3)

$$(1 + r)(1 + \frac{14}{5} \cdot \frac{1}{2}) + \frac{3}{5} w = 3$$

This means that

$$w = 1 - 4r$$

Thus the relation between w and r is linear.

If we divide equation (4.3) by 3 and (4.4) by 4 we obtain the relation expressing the production of 1 ton corn and 1 ton copper. We then find that the ratio of the labor values are $\frac{1}{2}$, i.e. as the price.

We furthermore find that

$$(1 + \frac{14}{5} \cdot \frac{1}{2}) \frac{5}{3} = (1 + \frac{6}{5} \cdot \frac{1}{2}) \frac{5}{2}$$

e.g. the same organic composition in the two production processes.

In Figure 1 we have drawn the factor-price curve for a specific technique α. Every point on the curve is determined by equation (4.2) above.

We measure production, using commodity A, and we can write the relation between net production per capita q, the wage rate w, and the profit rk in the following way

$$q = rk + w \tag{4.5}$$

The largest share that labor can obtain at the technique selected is w_{max} at $k = 0$. This means that we can write (4.5) in the following way

$$k = \frac{w_{max} - w}{r}$$

In the diagram this means that the tangent of the angle w_{max} pw measures the value of capital k.

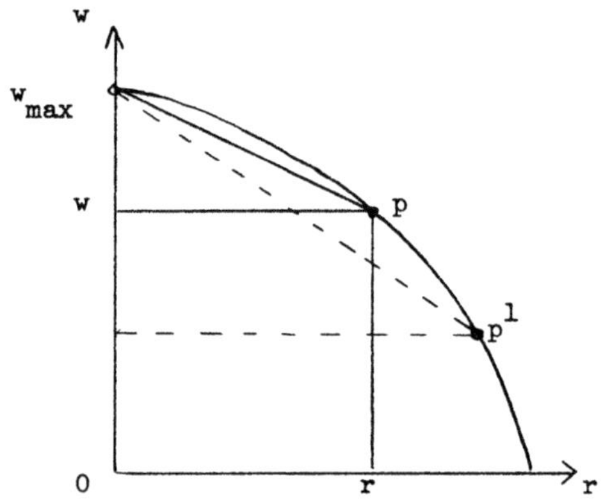

Figure 4.1

We can now also see from the figure that when r increases, then the value of capital goes up since the angle becomes larger.

To illustrate this further we shall study a factor-price diagram in which we have drawn two different techniques, namely α and β (see Figure 4.2).

We see that when the interest rate is zero, technique β is selected since this is the technique which then gives the highest wage per capita. At a there is a switch to the technique α. Capital per capita now goes down in the way shown in the forth quadrant from k_o to k_a. We see that the technique β has the highest capital intensity per capita since $\bar{w}_\beta a w_a > \bar{w}_\alpha a w_a$.

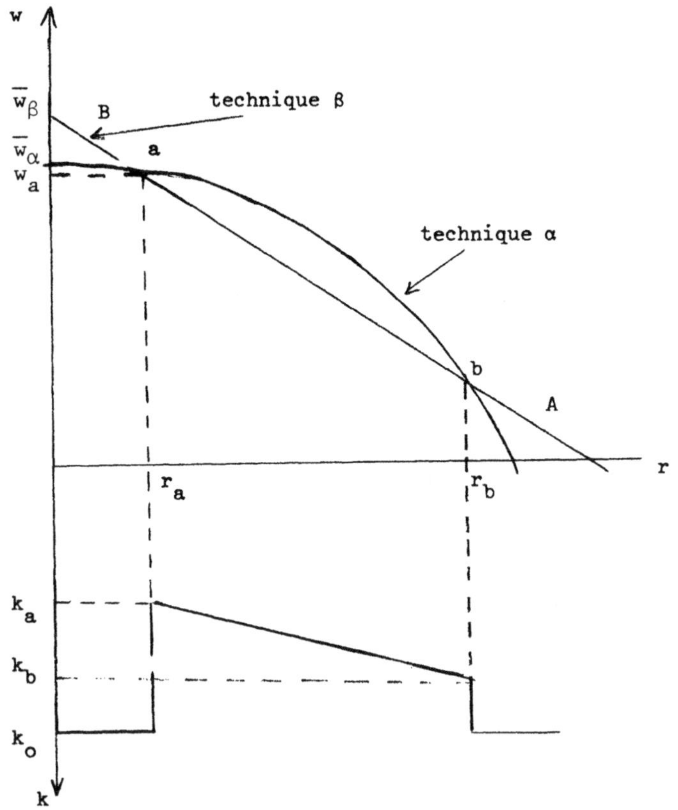

Figure 4.2

As the rate of interest goes up to b, "double switching" takes place and the technique β is again the one selected. We see in the lower part of Figure 4.2 that capital per capita is now increasing when the rate of interest goes up, which is the contrary to what was considered possible, as was mentioned above.

We have shown two possible consequences of the change of technique, namely

(1) "Double switching", which means that the same technique can be the best one at two different values of the rate of interest r.

(2) "Capital reversing", which means that capital moves in the same direction as the rate of interest. A lowering of the rate of interest implies that a less mechanized technique is chosen.

When there are only two techniques and when stationary states are compared, capital reversing and "double switching" always occur at the same time. If one, however, compare many techniques, one can have "capital reversing" without "double switching". A technique is then the best one over a certain interval and when it has disappeared it never comes back again. One important topic in the capital controversy has been the possibility to use the production function for the analysis of income distribution.

4.2 The production function and income distribution

If we have a production function which is homogenous of the first degree then then the function

$$Q = F(K, L)$$

where Q is quantity produced, K i capital, and L is labor, can be written

$$Q = LF(\frac{K}{L})$$

The wage rate w is assumed to be equal to the marginal productivity of labor, and the rate of profit is equal to the marginal productivity of capital, i.e. (we write the first and second derivatives as F' and F'' respectively)

$$w = \frac{\partial Q}{\partial L} = F\,(\frac{K}{L}) - \frac{K}{L}\,F^{1}(\frac{K}{L})$$

$$r = \frac{\partial Q}{\partial K} = F^{1}(\frac{K}{L})$$

$$\frac{dw}{d(\frac{K}{L})} = F^{1} - F^{1} - \frac{K}{L}\,F^{11} \qquad (4.6)$$

$$\frac{dr}{d(\frac{K}{L})} = F^{11} \qquad (4.7)$$

From (4.6) and (4.7) it follows that

$$\frac{dw}{dr} = -\frac{K}{L} \tag{4.8}$$

If we multiply (4.8) by $\frac{r}{w}$ we get

$$-\frac{dw}{dr}\,\frac{r}{w} = \frac{r \cdot K}{w \cdot L} \tag{4.9}$$

From (4.9) it follows that the wage profit-elasticity measures the distribution of income between the two factors of production.

4.3 The surrogate production function

Samuelson (1962) has tried to derive a production function which one should be able to use in spite of the possibilities of double switching and which makes it possible to use a production function for determining income distribution.

We shall describe how Samuelson constructs his production function since it illustrates the difficulties associated with such work if one takes into account what was discussed in Chapter 3.

Samuelson studies a simplified situation when one consumption and one capital good is produced using a given technique. The price of the consumption good is set equal to 1. The amount of capital and labor used in the consumption sector is k_c and ℓ_c, respectively. We obtain the following two equations (in the Ricardo-Sraffa system)

$$rp_k\, k_c + w\ell_c = 1$$

$$rp_k\, k_k + w\ell_k = p_k$$

If we eliminate p_k we obtain the following w - r relation

$$w = \frac{1 - rk_k}{r(k_c\ell_k - k_k\ell_c) + \ell_c} \tag{4.10}$$

Now Samuelson assumes that

$$k_c = k_k = k^*; \quad \ell_c = \ell_k = \ell^*$$

The consequences of these assumptions are that

$$p_k = p_c = 1$$

which means that (4.10) can be written

$$w = \frac{1 - rk^*}{\ell^*}$$

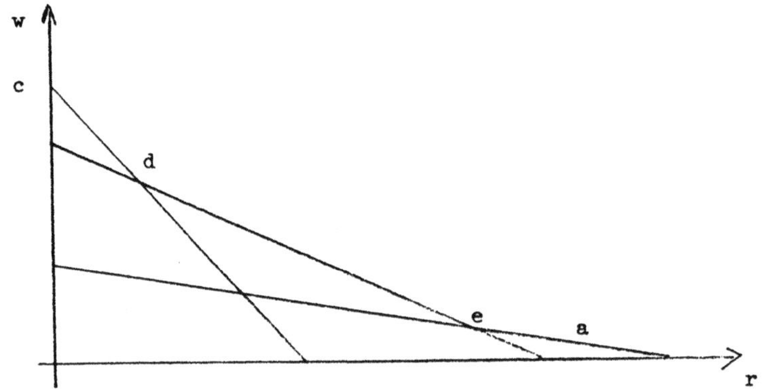

Figure 4.3

Figure 4.3 illustrates the factor-price lines for three different techniques. We can now draw the envelope to the factor price lines which is called the factor-price frontier. This gives us a production function which can be used to explain the income distribution.

By definition we have that

$$Y = Kr + Lw \tag{4.11}$$

Y = national income using the consumption good as numeraire

L = number of workers

r = rate of profit

w = wage rate

Dividing (4.11) by L we obtain

$$y = kr + w \tag{4.12}$$

Differentiating (4.12) we have

$$dy = rdk + kdr + dw \tag{4.13}$$

We now <u>assume</u> that

$$r = \frac{dy}{dk} \tag{4.14}$$

which, using (4.13), gives

$$k = -\frac{dw}{dr} \tag{4.15}$$

The production function constructed by Samuelson has the elasticity (since it is derived from the envelopeto linear factor-price curves)[1)]

$$e = -\frac{r}{w}\,\frac{dw}{dr} = \frac{rk}{w}$$

which gives the distribution of income. This means that we can explain income distribution in society using neo-classical theory.

1) If the factor price curve is linear we see (Figure 4.4) that what goes to labor is w and capital receives $w_{max} - w$. We can write $w_{max} - w$ as $-\frac{dw}{dr}\cdot r$ and therefore $\frac{w_{max} - w}{w} = -\frac{dw}{dr}\cdot\frac{r}{w}$ or the elasticity at the point p. If the production function is constructed by the envelope to linear factor price curves, we find that the elasticity measures income distribution at every point.

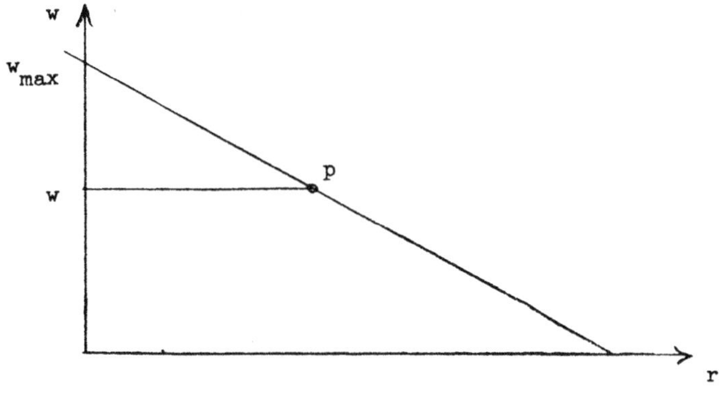

<u>Figure 4.4</u>

4.4 Critique of Samuelson's production function

We can see from (4.13) that the marginal productivity of capital $\frac{dy}{dk}$ is in general not equal to the rate of profit. In order for this to be true, it is necessary that $-\frac{dw}{dr} = k$, which we have to assume.

The consequences of this assumption can be illustrated using the figure below.

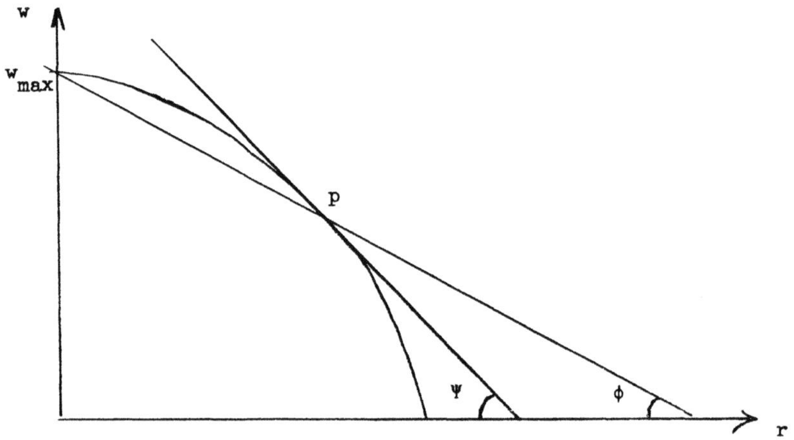

Figure 4.5

At the point p, $-\frac{dw}{dr} = k$ according to the assumption which Samuelson had to make. On the other hand we have by definition using (4.12) that $\frac{y - w}{r} = k$ and at the same time we have that $y = w_{max}$ when $r = 0$. It is therefore necessary that tan Ψ is the same as tan φ for Samuelson's results to be true. The assumptions made by Samuelson mean that the capital/labor ratio is the same in all industries, which in Marxian terminology implies an assumption about uniform organic composition of capital. For further discussions of this, see Bhaduri (1969).

If we do not introduce the restrictive assumption that the factor-price curves are linear but that they can have any form, it also follows that the envelope can have various shapes, e.g. as is shown in Figure 4.6.

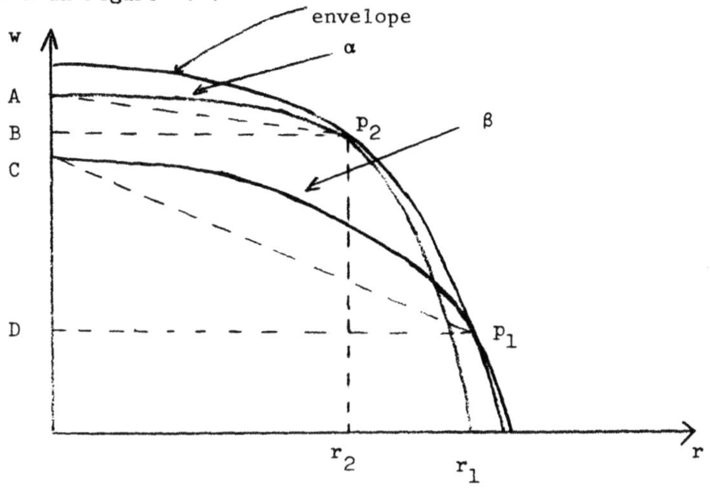

Figure 4.6

It follows that when r goes down, one will eventually go over from technique β to α and the value of capital goes down from Cp_1D to Ap_2B, i.e. when the rate of profit goes down, then the amount of capital used goes down. This is not supposed to happen according to neo-classical theory.

Garegagni (1970) has shown that it does not necessarily exist a unique relation between q and k. According to neo-classical theory, an increase of k will lead to an increase in q and for every value of k there is only one corresponding value of q. Garegagni has, however shown that the relation between q and r can be as is shown in Figure 4.7 which means that for a value of k e.g. k^* there exist two values of q.

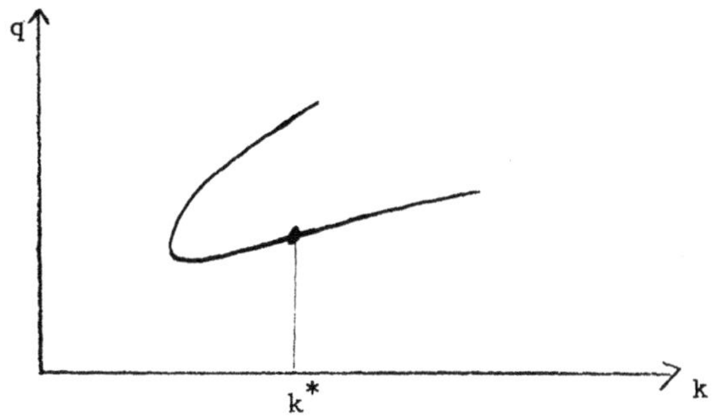

Figure 4.7

4.5 The importance of the rate of interest

We have pointed out many times that something is missing in order to "close" the economic system. Could this not be done by the rate of interest? In order for this to be possible we must give such a representation of the rate of interest that it can be determined independent of prices and wages in the economic system.

Fisher (1930) has given two different interpretations of the rate of interest, namely

(1) the rate of interest which makes two techniques equally profitable,

(2) the rate of interest which gives the value in all future periods of a sacrifice now.

The first interpretation need not give a unique rate of interest but if such a rate of interest exists, it is a pure definition which cannot explain anything. The other interpretation is more interesting and this is the one we shall mainly investigate in this section.

The neo-classical economists -- mainly Solow (1967) -- have tried to use the rate of interest to close the system. Solow uses the rate of interest as the link between micro- and macro-models of the economy. We shall here investigate what role the rate of interest can play.

4.5.1 One commodity

We begin by studying a society which produces only corn and which considers using one of two different techniques, α and β, which differ only with respect to the amount of corn being produced. Technique α gives y_{α} tons of corn every year, while β gives y_{β} tons.

Assume now that technique α is used but that one considers to begin using technique β. To be able to switch from α to β it is necessary to make an investment amounting to $y_\alpha - \bar{y}$, where $\bar{y}$ is the (reduced) quantity being consumed during the period when one goes from technique α to technique β.

The rate of interest obtained on the "investment" $y_\alpha - \bar{y}$ is defined as

$$\hat{r} = \frac{y_\beta - y_\alpha}{y_\alpha - \bar{y}} \qquad (4.16)$$

We see directly that the expression (4.16) has the desired properties of being independent of prices and the rate of profit. If the existing rate of profit is r and $r < \hat{r}$, where $\hat{r}$ is defined by (4.6), then one will switch to β.

We can now draw the following conclusions[1)]

(1) If $r < \hat{r}$ then that technique is most profitable which has the largest investment per capita in corn.

(2) If $r = \hat{r}$ then the two techniques are equally profitable and $\hat{r}$ is a measure of the first definition of the rate of interest r^* on page 4:12.

(3) The difference $r - \hat{r}$ which determines the choice of technique is a monotonic function of r since $\hat{r}$ is constant. This means that there is only one point at which there is a switch between techniques.

In the choice between techniques it is now reasonable to assume that one first uses the best method known (best land) and then one might have to utilize inferior qualities. If the number of techniques is infinite, we obtain a curve of the type shown in Figure 4.8.

---

1) r is the rate of profit which exists in the system exactly when one considers to make the investment. The variable r depends upon the prices in the system, but $\hat{r}$ does not.

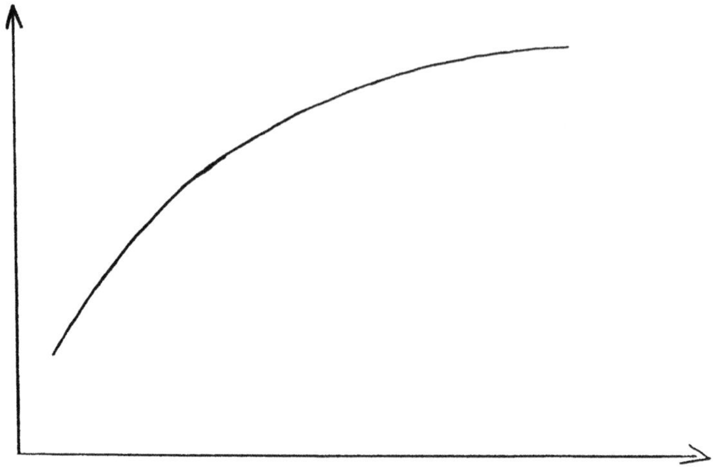

Figure 4.8

We now can see that the rate of interest emerges as a concept which exists only as a consequence of technical circumstances.

If a society has a given amount of corn, then we directly find the value of additional corn from the curve. Capital has a unique marginal value which is measured by the slope of the curve, which is independent of prices and wages. As capital goes up, the return on additional capital goes down.

What happens in a more realistic world with many commodities which change simultaneously?

4.5.2 Many commodities

To broaden the discussion in the previos section, we introduce an assumption which is crucial for the argument below, namely

(a) If -- at a certain wage rate w^*(and profit rate r^*) -- two techniques α and β are equally profitable, then at a different wage rate w the technique using more capital will be the most profitable when $w > w^*$ (or, which is the same thing, when the rate of profit $r < r^*$).

In addition we shall make the following two assumptions

(b) It is always possible for the economic system to go from one technique to another without loss of capital.

(c) There exists a very large number of techniques which can be ordered according to (a).

We shall use the following notations:

Y_β, Y_α = production using techniques β and α. Y is a vector with components Y_i (i=1,2,...,n) which represent commodity Y_i.

K_β, K_α = amount of capital when using technique β and technique α respectively. K is a vector (1,2,...,n).

$\bar{p}(r)$ = the price vector when the profit is r. $\bar{p}$ is a vector (1,2,...,n)

We can now formulate the rate of interest associated with a switch from technique α to β

$$\rho(r) = \frac{\bar{p}(r)(Y_\beta - Y_\alpha)}{\bar{p}(r)(K_\beta - K_\alpha)} \qquad (4.17)$$

Let us now consider the situation when we increase capital from K_α to K_β which increases the amount of goods from Y_α to Y_β. In order for this to be profitable it is necessary that (4.17) will produce a result such that $\rho > r$.

But now we use assumption (a) that $r < r^*$ where r^* is the level of profit at which the techniques α and β are equally profitable. In the same way we now have that $\rho < r$ for prices such that $r > r^*$ [1)].

---

1) Thus if $\rho > r$, the more capital intensive technique is selected.

What we have said so far means that ρ has all properties and behaves exactly in the same way as before, when we only had one commodity, i.e. when there were no valuation problems.

As before we now allow the number of techniques to be very large. At every level of profit there is always one technique which is the most profitable one. It also follows that for every alteration of the profit -- however small -- there will be a corresponding alteration of the most profitable technique. There are however some problems with this line of reasoning.

One problem with the argument above is that $K_\beta - K_\alpha$ does not need to be small only because the profit alters a little. If society uses a technique with a certain rate of return associated with it, the next technique, i.e. the one which increases the profit with a small amount may differ substantially from the one currently in use.

However, the decisive argument against the discussion above is that when double switching can take place, assumption (a) is no longer valid. This means that $r - \rho(r)$ can become zero, negative, or positive for any level of the profit, e.g. it is no longer a monotone function of the rate of profit. Consequently, the rate of return can no longer be used as a measure of the value of capital which depends upon technological conditions alone. For further discussions of the above problems, see Pasinetti (1969).

4.6 Conclusions

If double switching is possible, it is of great interest to investigate what the main consequences are for traditional economic theory.

The envelope of the factor-price curves associated with the different techniques will always decline, which also holds in traditional theory. If the rate of interest goes down and if the same technique is used, the workers can receive more wages, i.e. the factor-price curve is falling. Since the rate of interest has gone down there might, however, exist a still better technique, which would give the workers even more and thus increase the slope still more.

Double-switching has an important consequence for the relation between national income and the rate of interest. Earlier it was considered self-evident that a decrease of the rate of interest corresponded to an increase of national income. In order to enjoy a better life in the future it was necessary to make sacrifices now and in that process marginal productivity of investment and thereby the rate of interest would go down. If there is double switching, we can see from Figure 4.2 that we can now move from A to B without making any sacrifices at all. It is only necessary to reduce the rate of interest which, without alteration in the technique, increases the share of wages.

Traditionally, it was considered self-evident that the capital-output ratio would go up when the rate of interest goes down. This is, for instance, a consequence of using an ordinary Cobb-Douglas function. When there is double switching, it is possible that a lower share of profit is associated with a lower steady state capital income ratio.

The most striking consequence is that the same amount of capital gets different values, depending upon at which rate of profit it is used. A certain technique (amount of "capital") can be used at more than one profit and thereby at different prices. But then one does not know how to measure that "capital".

CHAPTER 5 INCOME DISTRIBUTION AND GROWTH

It followed from the discussions in Chapter 2 that four main problems were associated with Ricardo's theory, namely

(1) There was no satisfactory solution of the income distribution problem. This problem was solved by Sraffa as was discussed in Chapter 3.

(2) Ricardo did not completely understand problems associated with effective demand, as did Malthus, but not until the great depression was the problem clearly formulated and generally understood.

(3) The wage rate was predicted to remain at the subsistance level since higher wages would mean that the workers got more children. Ricardo had no explanation of and increased standard of living among the workers.

(4) There was a definite limit to population growth. A new human being brought two new hands to the world but they had to be employed on less fertile soil than the previous ones. Ricardo had no explanation of technological development and population growth.

As far as effective demand is concerned, Malthus attacked the opinion that every new human being is a benefit to society and he stressed that if production is much larger than consumption then we must reach a stage where the willingness to save and produce ceases because of the lack of sufficient consumption.[1)]

The basic principle which Ricardo did not understand was the difference between a primitive economy where every peasant and artisan produces what he can and want to produce. What is left he takes to the market where he sells it for whatever price he can get.

---

1) Malthus, T.R., Principles of Political Economy Considered with a View to their Practical Application, London 1820.

In a modern capitalistic economy things do not happen this way. Prices and wages are rigid and corporations react on changes in demand largely by altering volume. Demand determines how large production will be. If demand goes down then buildings, machines, and workers will have to remain idle. This causes further decline of demand and a depression develops. The important thing is to keep demand at a high level which was Keynes' message.

Concerning the theory of the growth of population and diminishing return, Ricardo's problem was more influenced by historical factors. In those days land was much more important than it is now and when land became less important some "failures" of the Ricardian theory were due to the followers of Ricardo who extended the concept of marginal rate of return to "capital". The only thing necessary to make Ricardo's theory function was to incorporate technological development, i.e. instead of assuming that marginal return is declining over time, it is possible that production is always rising due to technological development.

5.1 Conditions for long run growth

If we accept the central role of effective demand, the problem however is what determines it and how it can be sustained over time. Domar (1947), like Keynes, assumed that the willingness to invest, I, was decisive and that if the propensity to save, s, is known then demand, Y, is determined by

$$\frac{I}{s} = Y \tag{5.1}$$

but if investment is increased then total capacity will go up

$$\frac{IY}{K} = \frac{dQ}{dt} \tag{5.2}$$

where Q is production capacity and K is capital.

The important thing now is to investigate the conditions under which effective demand and production capacity grow at the same pace.

If we start from a hypothetical initial state, effective demand and production capacity must coincide and futhermore the growth of the capacity and the growth of demand must be equally large which means that the following differential equation must hold

$$\frac{dQ}{dt} = \frac{dY}{dt} \qquad Q(o) = Y(o) \tag{5.3}$$

If we insert (5.1) and (5.2) in (5.3) and use $I = sY$ we obtain

$$\frac{IY}{K} = \frac{dI}{dt} \cdot \frac{1}{s} \tag{5.4}$$

Let k be the capital output ratio and solve (5.4), which gives

$$I(t) = I(o) \cdot e^{\frac{s}{k}t} \tag{5.5}$$

This means that we must have exponential growth in order to maintain full employment and the rate of growth g is determined by

$$g = \frac{s}{k} \tag{5.6}$$

The problem with (5.6) is that one cannot guarantee that the nation can sustain the rate of growth g. If g is very large it may not be possible to achieve, given the rate of growth of the population and of technology. In the long run g must be equal to the natural growth which is composed of technological development and population growth i.e.

$$g_n = g \tag{5.7}$$

where g_n i the rate of natural growth which is composed of the rate of growth of the population, n, and technological growth, λ. The natural rate of growth does not have to be exponential but this is something which we <u>assume</u>.

Combining (5.6) and (5.7) we obtain

$$s = g_n \; k \tag{5.8}$$

By definition we have

$$s = \frac{S}{Y}$$

which means that (5.8) can be written

$$\frac{s_w \; W + s_c \; P}{Y} = g_n \; k \tag{5.9}$$

where s_w is the workers' propensity to save; s_c is the capitalits' propensity to save; W is total wage and P is the share of national income going to the capitalists.

Condition (5.9) determines the distribution of income and it closes systems like Sraffa's. If we make the usual assumption that $s_w = 0$ equation (5.9) takes the form

$$s_c \frac{P}{Y} = g_n \; k \tag{5.10}$$

We now see that the share of national income which goes to capital is determined. What goes to the workers is then the residual. It should be observed that by introducing technological development, population growth, and effective demand we have arrived at conclusions directly opposite to what Ricardo found. It is the share of capital that is determined by (5.10) and not the share going to labor by the iron law (see Fig. 2.1). Ricardo thought that it was the capitalists who absorbed the surplus once the workers had received their share determined by nature. Here the share going to capital is determined by nature (population growth and human inventiveness) and what remains goes to the workers.

As was shown in equation (5.8) the following relation must hold

$$s = g_n k \tag{5.8a}$$

We have mentioned above that g_n is exogenously given. Equation (5.8a) can be made to hold if s and/or k is variable. One dividing line between neo-classical and post-Keynesian (neo-Ricardian) theory concerns which of s or k is variable.

We have indicated the view which will be developed further, namely that s is variable while k is constant. We shall show that this can be used as the mechanism which closes Sraffa's system. Before proceeding we shall, however, indicate the neo-classical view according to which k is variable and s is constant.

5.2 The neo-classical view

In neo-classical theory it is assumed that there exists a production function f which depends upon capital K and labor L, i.e. we have

$$Y = f(K,L) \tag{5.11}$$

The function f is assumed to be differentiable with positive first derivatives and negative second derivatives.

If there is perfect competition in both labor and capital markets, then the wage rate as well as the rate of interest will be determined by the marginal productivity of each factor of production. Full employment is assumed to prevail and the labor force to grow with an exogenously given velocity $L = L_o e^{nt}$. The growth of capital is determined by

$$\dot{K} = s\, f(K,L) \tag{5.12}$$

Since L is given from the full employment requirement, we can principally solve K from (5.12). This can then be inserted in (5.11) and we can solve for Y as a function of time (since both K and L are known).

It can then be shown that the solution of (5.12) is stable and that the demand and supply conditions are adjusted to fulfill the Harrod-Domar conditions.

If we, as an illustration, assume that the production function is Cobb-Douglas, we have that (use (5.12))

$$\dot{K} = sK^{\alpha} \cdot L^{1-\alpha} = sK^{\alpha} \cdot L_o^{1-\alpha} \cdot e^{n(1-\alpha)t} \tag{5.13}$$

Assume that we know the initial value of capital $K(o) = K_o$. We are then able to solve the differential equation (5.13) and obtain

$$K(t) = \left[K_o^{1-\alpha} - \frac{s}{n} L_o^{1-\alpha} + \frac{s}{n} L_o^{1-\alpha} e^{n(1-\alpha)t}\right]^{\frac{1}{1-\alpha}} \tag{5.14}$$

If we now insert (5.14) in the production function $Y = K^{\alpha}L^{1-\alpha}$ and write the capital output ratio $k = \frac{K}{Y}$, we obtain

$$k = \left[\frac{K_o}{L_o} e^{-nt}\right]^{1-\alpha} - \frac{s}{n} e^{-n(1-\alpha)t} + \frac{s}{n} \tag{5.15}$$

When t gets large, k approaches $\frac{s}{n}$, i.e. k has adjusted in such a way that the Harrod-Domar conditions are fulfilled.

At this point, we shall not penetrate the neo-classical analysis any further (the main criticism is the large number of restrictive assumptions which are necessary) but the reader is referred to e.g. Harcourt (1972), Solow (1956), and Pasinetti (1974) for more detailed discussions.

5.3 A combined view

Since we have argued that the two different schools of thought being investigated have different views about whether it is s or k that is variable in equation (5.8a), it might perhaps be of some interest to investigate what happens if both s and k are variable (as Pasinetti (1974) has done).

If we set $s_w = 0$, we can write (5.9) as

$$g_n \frac{K}{Y} = s_c \frac{P}{Y}$$

We now see that regardless of whether k varies of not we must have that

$$\frac{P}{K} = \frac{1}{s_c} g_n$$

which is the post-Kenyesian result (i.e. the one we previously arrived at by variations of s).

With the introduction to post-Keynesian distribution theory as a background, we shall discuss its development. We use the names of those persons who mainly contributed to the development of the field as titles of the sections to follow.

5.4 Kaldor

In the previous chapters we have shown that the economic system developed by Sraffa needs something from outside to be closed. We shall here introduce two things which are assumed to be independent of the economic system itself, namely

(1) the society is divided into two social classes, workers and capitalists, with known propensities to save. This assumption will be dropped in the next section
(2) the economic growth is known and might, for example, depend upon population growth and technological growth.

To begin with, we shall associate the propensity to save s_w with labor and s_c with the capitalists. Later we shall show that more general classifications can be used.

It follows from (5.10) that investment is determined by conditions for long run growth. In the models discussed here, one need not necessarily refer to long run growth as a determinant of investment, but rather take investment as exogenously given by e.g. "the animal spirit of the capitalists. We shall always assume that the economy is in a state of full employment.

The following notations will be used

S = total saving

P = total profit

W = total wage

I = investment

Y = income

The following relations must hold

$$S = s_c P + s_w W = I \tag{5.16}$$

$$P + W = Y \tag{5.17}$$

We can write (5.16) in the following way using (5.17)

$$s_c P + s_w (Y - P) = I$$

which can be written

$$\frac{P}{Y} = \frac{1}{s_c - s_w} \frac{I}{Y} - \frac{s_w}{s_c - s_w} \tag{5.18}$$

We have assumed that $\frac{I}{Y}$ is constant and known. This ratio will now together with s_w and s_c determine the distribution of income as is shown by (5.18). Equation (5.18) will hold within an interval where $\frac{I}{Y}$ is not so high that the workers will make a revolution or not so low that willingness to invest is destroyed.

If we set $s_w = 0$ in (5.18) we obtain

$$\frac{P}{Y} = \frac{1}{s_c} \frac{I}{Y}$$

and we see that the capitalist class can determine its share by varying s_c. The smaller the s_c, i.e. the less the capitalists save, the larger their share of national income. Note, however, that this refers to their share and not to what their absolute income would be.

If the economy is growing in such a way that capital and total income grow with the same velocity, it holds that

$$\frac{I}{Y} = \frac{\Delta K}{K}\frac{K}{Y} = g\frac{K}{Y} \qquad (5.19)$$

where K is capital and g is the rate of growth $(g = \frac{\Delta Y}{Y} = \frac{\Delta K}{K})$.

We can now write the rate of return on capital in the following way

$$r = \frac{P}{K} = \frac{P}{Y}\frac{Y}{K} = \frac{Y}{K}\left[\frac{1}{s_c - s_w}\frac{gK}{Y} - \frac{s_w}{s_c - s_w}\right]$$

or

$$r = \frac{P}{K} = \frac{g_n}{s_c - s_w} - \frac{s_w}{s_c - s_w}\frac{Y}{K} \qquad (5.20)$$

If we set $s_w = 0$ we obtain

$$r = \frac{g}{s_c} \qquad (5.21)$$

and if $s_c \to 1$, then $r = g$, which is the result obtained by von Neumann.

5.5 Pasinetti

In the analysis so far we have not observed that if the workers save, then they also become owners of capital and therefore in addition to their wage receive a fraction of total profit.

We must now go back to equation (5.16) and divide the profit into two parts P_k which goes to the capitalists and P_w which goes to the workers. We have previously used s_c as the propensity to save out of profits which was also similar to the capitalists' propensity to save. In order to keep the propensity to save out of profits apart from the capitalists' propensity to save we denote the latter by s_K.

$$S = s_K \cdot P_K + s_w(W+P_w) = I \qquad (5.22)$$

$$P_K + P_w + W = Y$$

We can now proceed as in the previous paragraph and derive the equivalent of (5.18)

$$\frac{P_K}{Y} = \frac{1}{s_K - s_w} \frac{I}{Y} - \frac{s_w}{s_K - s_w} \tag{5.23}$$

We can rewrite (5.23) in the following way

$$\frac{P_K}{K} = \frac{1}{s_K - s_w} \cdot \frac{I}{K} - \frac{s_w}{s_K - s_w} \frac{Y}{K} \tag{5.24}$$

A certain fraction of capital is owned by the workers which we denote by K_w. They receive an income from capital equal to interest which the owners of capital pay for borrowing the workers' money. Assume that this rate of interest is i. The total share of income going to capital is $P + iK_w$ and we obtain, using (5.24)

$$\frac{P}{K} = \frac{1}{s_K - s_w} \frac{I}{K} - \frac{s_w}{s_K - s_w} \frac{Y}{K} + \frac{iK_w}{K} \tag{5.25}$$

In a dynamic equilibrium it must hold that

$$\frac{K_w}{K} = \frac{S_w}{S} = \frac{s_w (Y-P_k)}{I} \tag{5.26}$$

In the long run, the interest received by the workers on their capital must coincide with the return received by the capitalists which is the same as the return on all capital; otherwise only one social class would remain in the long run. We therefore obtain

$$i = \frac{P}{K} \tag{5.27}$$

If (5.26) and (5.27) are inserted in (5.25) we obtain, using (5.24)

$$\frac{P}{K}\left(1 - \frac{s_w s_K}{s_K - s_w} \frac{Y}{I} + \frac{s_w}{s_K - s_w}\right) = \frac{1}{s_K - s_w} \frac{I}{K} - \frac{s_w}{s_K - s_w} \frac{Y}{K}$$

i.e.

$$\frac{P}{K} \frac{s_K(I-s_wY)}{1} = \frac{I - s_wY}{K}$$

and if $I > s_w Y$

$$\frac{P}{K} = \frac{1}{s_K} \cdot \frac{I}{K} = \frac{g}{s_K}$$

Furthermore, since $gK = I$, we obtain

$$\frac{P}{Y} = \frac{1}{s_K} \frac{I}{Y}$$

Thus the distribution of income between wages and profits does not depend upon s_w.

One main observation from this analysis is that the owners of capital are assumed to be able to control their own accumulation of capital and investment which is an independent datum to which the saving by the workers is adjusting, i.e. $\frac{s_w(W+P_w)}{P_w} = \frac{S_w}{P_w}$ is adjusting. This means that if the workers save, then they receive such a profit that their saving out of both wages and capital equals the amount of saving which the capitalists should have contributed if they had received the profit going to the workers, i.e. if

$$s_w(W+P) = s_K \cdot P_w$$

5.6 Samuelson and Modigliani

When we consider the long run equilibrium and study the share of capital owned by the capitalists and by the workers, we shall show that one of the following situations can prevail

(1) there is no long run equilibrium determining the distribution of capital between the two social classes,

(2) the distribution of capital is such that $\frac{P}{K} = \frac{g_n}{s_K}$,

(3) the distribution of capital is such that $\frac{Y}{K} = \frac{g_n}{s_w}$.

We shall now illustrate how different values of the parameters determine which equilibrium situation a society will have.

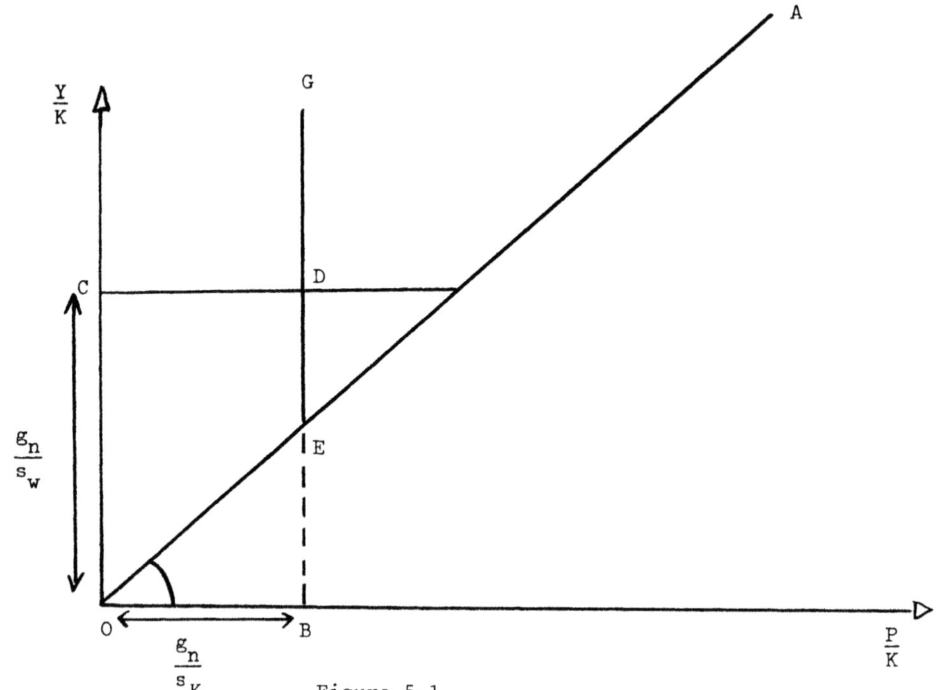

Figure 5.1

In Figure 5.1 we have put $\frac{P}{K}$ along the X-axis and $\frac{Y}{K}$ along the Y-axis. We have drawn the line OA through the origin having the slope 45°. The only feasable area in the figure is the one above OA since otherwise P > Y which is impossible.

We shall now prove that the only possible combinations of $\frac{Y}{K}$ and $\frac{P}{K}$ in the stationary state must either be located along CD or DE.

I. It is not possible to find any solutions to the right of BG since then $\frac{P}{K} > \frac{g_n}{s_K}$ which can be written $P \cdot \frac{s_K}{K} > g_n$. This means that the capital owned by the capitalists is growing faster than the natural rate of growth. This is not possible in the long run since this would mean that total capital would grow faster than g_n which is not possible.

II. Nor are solutions above CD possible since there we have that $s_w \frac{Y}{K} > g_n$. This means that if all receivers of income are wage earners, then capital will grow faster than the natural rate of growth g_n. If a share of capital is owned by the capitalists, capital will grow even faster.

III. Finally we must show that solutions inside OCDE also are impossible. Here we have both that $s_K \frac{P}{K} < g_n$ and that $s_w \frac{Y}{K} < g_n$. The condition $s_K \frac{P}{K} < g_n$ means that capital owned by the capitalists grows slower than g_n and thus it will finally disappear. Then all capital will be owned by the workers and $s_w \frac{Y}{K} < g_n$ which means that capital is growing at a rate lower than g_n which is impossible.

Along CD we have that $s_w \frac{Y}{k} = g_n$ and $s_K \frac{P}{K} < g_n$, i.e. all capital will be transferred to the workers. This is the case when capital is redistributed without revolution. This is duscussed by Samuelson-Modigliani but which cannot be accepted by post-Keynesians. If all capital is owned by labor then a mechanism for maintaining long run equilibrium based upon a variable propensity to save in society is not possible. As was discussed in Section 5.1, this is central to the post-Keynesian position.

Along DE it must hold that $\frac{P}{K} = \frac{g_n}{s_K}$, and now it will be possible for both capitalists and workers to exist with different propensities to save. For further discussions of this, see Samuelson-Modigliani (1966) and Meade (1966).

5.7 Kaldor again

An important starting point for the previous analysis has been that it is possible to divide society into two social classes with different propensities to save. The difference in propensity to save has been explained by the great wealth owned by the capitalists. They are very rich and therefore they can save a large fraction of their income.

It may, however, be more reasonable to start a discussion about saving out of capital from the situation of corporations. They must retain a substantial part of profits in order to finance product development, investment, and other activities which are necessary to guarantee the growth and development of the enterprise.

The corporation thus first decides how much it wants to retain of the profit and then distributes dividends to workers that own capital and to capitalists. In addition to the dividend given to the two social classes, they receive capital gains or losses and these gains and losses depend upon the evaluation of corporations by the stock market.

By bringing in the stock market we obtain two conditions for equilibrium, namely

(a) saving must be equal to investment, and

(b) prices in the stock market are such that demand for shares equals supply.

In order to analyze this further we shall use the following notations.

Total wage = W

Propensity to save from wages = s_W

The capitalists' net consumption from capital gains G is cG.

Corporations reinvest a fraction s_K of the profit

Corporations issue new shares to finance a fraction, i, of their growth. Capital = K; rate of growth = g, and therefore the value of newly emitted shares is igK.

Condition (b) can be written

$$s_w W = cG + igK \tag{5.28}$$

The value of existing shares is pN, where p is the price of shares and N denotes the number of shares. The increase in value of the shares is

$$G = N\Delta p = v\Delta K - p\Delta N$$

where v is the ratio of market value of shares pN and capital.

We have that

$$\Delta K = gK$$

$$p\Delta N = igK$$

$$G = vgK - igK \tag{5.29}$$

If we insert (5.29) in (5.28) we obtain

$$s_w W - c(vgK-igK) = igK \tag{5.30}$$

Condition (a) above requires that

$$s_w W - c(vgK-igK) + s P = gK \tag{5.31}$$

The idea above is that workers have a net propensity to save $s_w W$, e.g. in the form of pension funds. The supply of new shares come from capitalists cG and the new ones issued by companies igK. We now seek a valuation ratio v such that conditions (5.29) and (5.31) hold simultaneously.

Total profit $P = \rho K$ where ρ is the rate of profit and we know that $Y - P = W$. Thus we have that

$$s_w Y - s_w \rho K - cvgK + sigK = igK$$

and

$$s_w Y + (s_K - s_w)\rho K - cvgK + cigK = gK$$

We can now rearrange and divide by gK which gives

$$\frac{s_w}{g}\frac{Y}{K} - \frac{s_w\rho}{g} - cv + ci = i$$

$$\frac{s_w}{g}\frac{Y}{K} + \frac{(s_K-s_w)\rho}{g} - cv + ci = 1$$

If this is solved with respect to v and ρ, we obtain

$$v = \frac{1}{c}\left(\frac{s_w}{g}\frac{Y}{K} - \frac{s_w}{s_K}(1-i) - i(1-c)\right) \tag{5.32}$$

$$\rho = \frac{g(1-i)}{s_K}$$

It is interesting to note that the rate of profit ρ only depends upon g, i, and s_K, i.e. it is independent of s_w.

We have here obtained a useful link between the micro-conditions facing corporations on the stock market and conditions necessary for macro-economic equilibrium to which we shall return in Chapter 8.

5.8 Conclusions

In Chapter 3 we discussed various ways of closing the Sraffa system. In this chapter we have shown that it is possible - at least in principle - to use the Harrod-Domar condition and let the distribution of income between labor and capital be determined such that the Harrod-Domar condition is fulfilled.

Thereby we obtain one more equation and the Sraffa system can be closed. We also by this build in a new dynamic element in Ricardo's theory, which enables it to deal with societies where labor gets a larger share of income than what is determined by the subsistence level.

PART II

APPLICATIONS

CHAPTER 6 APPLICATIONS: AN INTRODUCTION

In his analysis of Marx' works, Balibar (Althusser and Balibar, 1970, p.209ff) finds that every mode of production consists of the following elements namely the labourer (labour power), the means of production (object of labour and means of labour), and the non-labourer.

Balibar also finds two different kinds of connexions or relations between the elements mentioned. The first relation is to be found on the social level and is a property-connexion. The second relation is a technical relation and is called "real or material appropriation connexion" by Balibar. This relation we can interpret as "putting the means of production into work".

In a capitalistic system of the kind described in Chapter 2, we find two different non-workers, the land-owner and the capitalist. The land-owner gets a rent and he does not take part in the work or in organizing the work. The capitalist owns the means of production, with the exception of land, and he organizes the production process. The worker, finally, owns his labour power only and he is also the one who puts the means of production into work.

In a more developed capitalistic society, like the one described in Sraffa's system (Chapter 3), the landowner has disappeared. In Sraffa's system we find means of production, profit and wages. One interpretation is that capitalists own the means of production and get the profits. In the Sraffa-system the interdependences in the production process are emphasized. Products are produced by means of other products. In a neo-classical system, like Walras', the interdependences in consumption are emphasized. In Walras' system the individual is the basic unit and we face an atomistic analysis. In the Sraffa-system we face a vertical analysis, the society is divided into two classes, capitalists and workers.

To some extent it is possible to say that the differences mentioned above have to do with the realism of different economic models. Are the basic units and relations in the Sraffa-system more relevant than those in a Walras-system? The answer to that question is dependent on what we consider as relevant. However, in one sense all equilibrium models are unrealistic. "Long-run equilibrium" is,as Joan Robinson has pointed out, not some future date but a logical device. Fruitfulness seems to be a more relevant criterium than realism. What kind of questions can we get answers to by using one or another equilibrium model? How the knowledge from an equilibrium model is to be used in real life is then another question. What we have said above is true for growth models too.

Of course it is also possible to evaluate different models and theories from a logical point of view. One example has already been given in Chapter 4, when it was shown that the marginalistic distribution theory was inconsistent. In Chapter 9, which deals with technological change, we give another example. Using neo-Ricardian concepts and methods, we show that the neo-classical measure of technological change, where one tries to distinguish between movements along a production function and shifts in a production function, is wrong.

To some readers, maybe, this emphasis on concepts and logical consistency of different theories seems strange, at least when unrealistic equilibrium models are in the focus. We believe, however, that even if we cannot apply theories directly, they influence us when we try to understand the real world. In that sense concepts and theories, in an indirect way, are of great practical importance.

In Chapter 9 we also describe a growth model (which considers technological change) based on Ricardian concepts. However, we start in Chapter 7 with a short discussion of some stability problems. The purpose is <u>not</u> to present a comprehensive treatment of these problems. Instead the purpose is to outline how the Sraffa-system gives <u>one</u> framework in analyzing stability problems and where one emphasizes the fight between capital and labour.

In Chapter 8, we deal with what we in the beginning of this chapter called the property-connexion. Using a model by Kaldor (presented in

Chapter 5) we analyze some capital sharing policies which have been discussed in Sweden and other countries.

In Chapter 10, we deal with the efficiency of different economic system. We use the concept double-switching when dealing with the efficiency problem. In this analysis, we face both of the two relations we mentioned in the beginning of this chapter, the social and the technical relation.

Finally, we want to emphasize that the purpose of the following chapters is not to give a thorough treatment of the problems we discuss. Rather, the purpose is to show that neo-Ricardian concepts and methods can be used when analyzing different economic problems. The neo-Ricardian theory is not only a critique of neo-classical theory, it has a positive content too.

CHAPTER 7 SOME STABILITY PROBLEMS: AN OUTLINE

7.1 Introduction

In this chapter we are going to discuss three problems which are related to effects of changes in wages, productivity and demand. We will discuss these problems keeping some aspects of inflation in mind. The concept "inflation" can be interpreted in at least two different ways. First, it can be interpreted as a cause of a general rise in prices and second as a synonym for a general rise in prices. It is not always clear which interpretation different authors use. However, when it comes to explaining inflation, it is easier to distinguish between different "schools". At least two "schools" can be found.

First we have a group of economists emphasizing cost-push inflation. According to this group inflation is caused by trade unions when they demand excessive wage increases, or it is caused by firms when they increase the prices of raw materials and other producers' goods in order to increase the profit rate.

Another group of economists emphasize demand-pull inflation. These economists often blame government for spending too much and taxing too little. Sometimes these two views, cost-push and demand-pull, are linked together and then we get the wage-profit-price spiral. (For a more detailed discussion about the two views, see Machlup, 1960.)

We are going to use Sraffa's model in our discussion. This may be seen as unusual because money does not exist in Sraffa's model. Without any doubt we miss one important factor when we try to analyze stability problems by utilizing a model which does not include money.

However, one problem with traditional analysis is, according to Yeager (1968), that money serves as both unit of account and medium of exchange. In a footnote he points towards "the farfetched but theoretically challenging concept of a system in which the two functions are split, with the actual medium of exchange fluctuating in price in terms of the separate unit in which ordinary goods and services are also priced". By using Sraffa's model with its numeraire, the standard commodity, we have a "separatare unit" of the type Yeager is looking for.

It should also be pointed out that we compare equilibrium points, that is to say time is not included in the models. But as indicated already in the title to this chapter, the purpose is not to give a comprehensive treatment of stability problems. Rather, the purpose is, by using numerical examples, to indicate how stability problems can be looked upon, using the methods presented earlier in this book.

7.2 Effects of a wage increase

In this section we are studying the effects of a wage increase keeping the discussion about cost inflation in mind. We assume strong labour unions which fight for a larger share of the total national product. In this section we disregard productivity changes.

Let us start with the example we utilized in Chapter 3. We modify that example so we get a third sector producing a non-basic product. We assume the following economic system:

$$\begin{aligned} 2A + 2B + \frac{18}{25} L &\rightarrow 8A \\ 2A + 5B + \frac{6}{25} L &\rightarrow 8B \\ A + B + \frac{1}{25} L &\rightarrow 8C \end{aligned} \qquad (7.1)$$

The system has two basic products A and B and one non-basic product C (L stands for labour). Observe that C is not used as an input in the C-sector. In Appendix A at the end of this chapter, we show what can happen if a non-basic product is used as input.

In order to compute the prices we formulate (7.1) in the following way:

$$\begin{cases} (2p_a+2p_b)(1+r) + \frac{18}{25} w = 8p_a \\ (2p_a+5p_b)(1+r) + \frac{6}{25} w = 8p_b \\ (\ p_a+\ p_b)(1+r) + \frac{1}{25} w = 8p_c \end{cases} \qquad (7.2)$$

R, the maximum profit rate, is equal to $\frac{1}{33}$(as it was in (3,13)). The standard product in physical terms is

$$\frac{10}{6} \text{ units of A}$$
$$\frac{20}{6} \text{ units of B} \qquad (7.3)$$

Observe that nothing of C is included in the standard product. This is always true for non-basic products (however, cf. Appendix A). In Table 7.1 is given the product prices for different w-values. Prices and wage have the standard national income as the numeraire; a numeraire which is independent of the income distribution. V stands for the value of the gross product 8A + 8B + 8C.

w	p_a	p_b	p_c	V
1	0,18857	0,20571	0,05428	3,58848
1/2	0,15512	0,22240	0,05756	3,48064
1/4	0,13777	0,23111	0,05888	3,42208
0	0,12000	0,24000	0,06000	3,36000

Table 7.1

In Table 7.1 we see that when the wage is increasing then the price of A is increasing and the prices of B and C are decreasing. The value of the gross product V increases when the wage increases. From the following example we can see that this latter state of affairs is not always true.

$$\begin{aligned}(2p_a+2p_b)(1+r) + \tfrac{18}{25}w &= 8p_a\\ (2p_a+5p_b)(1+r) + \tfrac{6}{25}w &= 8p_b \qquad (7.4)\\ (\ p_b+4p_c)(1+r) + \tfrac{1}{25}w &= 8p_c\end{aligned}$$

In comparison with (7.2) the C-sector has been changed in (7.4). The gross product is still 8A + 8B + 8C and we get prices and V-value as shown in Table 7.2.

w	p_a	p_b	p_c	V
1	0,18857	0,20571	0,06143	3,64568
0	0,12000	0,24000	0,12000	3,84000

Table 7.2

That some prices increase and some decrease when the wage share increases has been known for a long time. Already Ricardo in his argument with Adam Smith about the origin of value, found that this could happen. (In Appendix B of this chapter, we have enlarged the analysis to a n-sector economy.)

A result of the example shown above (and the more general analysis in Appendix B) is that an increase in wages must not always be associated with an increase in the general price. This is now assumed automatically to be the case. When the economy is operating at full capacity and profits and the level of employment is high then labour is told to hold back wage demands because they will cause prices to go up. It is suggested instead that demands for higher wage levels should come when business is depressed. This can only be interpreted as a resistance against changes in the income distribution.

The reason why a wage increase is followed by an increase in the general price level could be that prices are flexible upwards but not downwards. This is a phenomenon which has been observed by many authors (e.g. Schultz, 1969, p.215). This means that changes in the income distribution are neutralized and that the general price level increases. In our case it means that the numeraire in the Sraffa-system breaks down. The conclusions from our analysis are the opposite of Paish's (1969, p.226). He says that there is no need to consider in detail any proposal for the direct limitation of profits. From our analysis it seems that one has to limit the profits at least if one wants the workers to have a larger share of the national product.

7.3 Effects of a productivity increase

In Section 7.2 we mentioned how the general price level increases when the workers want a larger share of the total product. The reason was that prices are not flexible downwards. In this section we want to study if one faces the same problem if the system has an increase in productivity. The two schools of inflation, cost-push and demand-pull, seem to agree on one point, namely that rises in productivity are anti-inflationary (see Streeten, 1969, p.181).

Once again we will use the example that we used in the preceding sections. In the following we assume that a productivity rise means that we need less input to produce a given output. We assume the following system.

$$
\begin{aligned}
(p_a+2p_b)(1+r) + \frac{18}{25}w &= 8p_a \\
(2p_a + 4p_b)(1+r) + \frac{6}{25}w &= 8p_b \\
(p_a+p_b)(1+r) + \frac{1}{25}w &= 8p_c
\end{aligned}
\qquad (7.5)
$$

In (7.5) R = 3/5. It can also be shown that the standard product in physical terms is

$$
\begin{aligned}
&\frac{15}{6} \text{ units of A} \\
&\frac{30}{6} \text{ units of B}
\end{aligned}
\qquad (7.6)
$$

which should be compared with (7.3).

The solutions to (7.5) for two different w-values are shown in Table 7.5.

w	p_a	p_b	p_c
1	0,14000	0,13000	0,03875
1/2	0,11000	0,14500	0,04393

Table 7.5

From Table 7.5 it is clear that the prices have decreased after the increase in productivity (cf. Table 7.1, where R = 1/3).

Generally, it is true that a productivity rise, as we interpreted it above, in the system of basic products gives an increase in R, the maximum profit rate. This can be shown utilizing Fröbenius' theorem (see Section 3.6).

One problem which we face when studying a productivity rise among the basic products is that the numeraire, the standard product, is changing (cf. (7.3) and (7.6)). However, we know that after an increase in productivity the surplus is larger. For a given income distribution more commodities will be allocated to the workers after the productivity rise.

It is sometimes said that the wage demands of the workers should not exceed the increase in productivity. This can be given one interpretation, which means that the workers have to bargain in order to keep the income distribution between labour and capital intact.

7.4 Effects of changes in demand[1)]

We start with the example which we utilized in the beginning of this chapter (cf. (7.2)).

$$\begin{aligned}(2p_a+2p_b)(1+r) + \frac{18}{25}w &= 8p_a\\ (2p_a+5p_b)(1+r) + \frac{6}{25}w &= 8p_b\\ (p_a+p_b)(1+r) + \frac{1}{25}w &= 8p_c\end{aligned} \tag{7.7}$$

For $w = 0{,}5$ we got the following solution

$$\begin{aligned}p_a &= 0{,}15512\\ p_b &= 0{,}22240\\ p_c &= 0{,}05756\end{aligned} \tag{7.8}$$

The net product in (7.7) is 3A and 8C when the value is 0,92944 using the standard commodity as a numeraire. The workers' share is 0,5, then the share of the capitalists must be equal to 0,42944. We assume that the system is in equilibrium.

Let us now assume that a change in the preferences occurs, so that the workers want to use their wages to buy product C to a value of 0,2 and product A to a value of 0,3. Furthermore we assume that the capitalists demand product C to a value of $154/1025 \approx 0{,}15024$. The consequence is that we will have a demand of 6 units of C, in other words we have a surplus of 2 units of product C.

1) This section is inspired by Bose (1965).

The net product of A is 3 units, its value equal to $954/2050 \approx 0{,}46536$. The workers want to spend 0,3 in buying product A. The capitalists want to spend 0,27561 in buying A (what is left of the profit when they have bought product C). So, after the change in preferences we have an excess demand of product A (the value of this excess demand is $226/2050 \approx 0{,}11024$, which is equal to 113/159 units of A when we use the prices in (7.8)).

If we assume that prices are flexible upwards but not downwards, then we also know that the system (7.7) is out of gear. If we have a market for product A which is totally free, then the whole system is in danger because we run the risk that workers and capitalists consume out of units of A which are needed in the production process. Even if we assume separated markets so the reinvestment program is guaranteed, it is not easy to see how the excess demand can be eliminated with taxes only.

The question is if the system can be brought back to equilibrium by moving resources from the C-sector to the A-sector through a market process or through some planning authority.

We assume the new production process to be determined by

$$\frac{4}{5} A + \frac{3}{4} B + \frac{4}{125} L \rightarrow 6C \qquad (7.9)$$

which means that we have no assumption about constant returns to scale. Furthermore, let us assume that the surplus of labour and resources (A and B) is transferred to the A-sector. Then we get the following system

$$\begin{aligned} 2\frac{1}{5} A + 2\frac{1}{4} B + \frac{91}{125} L &\rightarrow ?A \\ 2A + 5B + \frac{6}{25} L &\rightarrow 8B \qquad (7.10) \\ \frac{4}{5} A + \frac{3}{4} B + \frac{4}{125} L &\rightarrow 6C \end{aligned}$$

Even if this system produces 8 113/159 units of A, it cannot be taken for granted that system (7.10) will produce A to a <u>value</u> which is equal to the value demanded. This is dependent on the production relations. It should also be observed that R, the maximum profit rate, has been changed. The reason is that the production equation for a basic commodity has been changed. Therefore, we agree with Bose (1965, p.779) when he says:

> "Sectoral losses and gains will throw the whole system out of gear, instead of promoting corrective price-output adjustments. This is because the ultimate determinants of prices - the physical production relations and social wage-profit division - will not be "adapted" easily to the "force of demand" acting via divergent profit-rates, although they will almost certainly be changing autonomously from one time period to another."

It also seems to be the case that the difficulties of predicting effects of changes will increase the more the different sectors are connected to each other, in other words the more developed the economy is.

The policy measure described above, i.e. transferring resources from a stagnating sector to an expanding one, is related to the model of Rehn (see Andersson and Meidner, 1973, p.20ff). Even this measure will give instability problems. In Bose (1965), other policy measures are discussed.

7.5 Summary

In this chapter we have discussed some stability problems. Using numerical examples we outlined how Sraffa's model can be used when discussing these problems. The most important result was presented in Section 7.4 where we saw that demand factors, like preferences, can throw the whole system out of gear. It should however be pointed out that the analysis was based on a comparison between equilibrium points but not on a study of transitions between equilibrium points (which is the proper approach).

APPENDIX A

The purpose of this appendix is to show what can happen with the standard system if a non-basic product is used as input in its own production.

Let us start with the following production system.

$$2A + 2B + \frac{18}{25} L \rightarrow 8A$$

$$2A + 5B + \frac{6}{25} L \rightarrow 8B$$

$$B + cC + \frac{1}{25} L \rightarrow 8C$$

where c is an input parameter.

The matrix A (cf. Chapter 3) is

$$A = \begin{pmatrix} \frac{1}{4} & \frac{1}{4} & 0 \\ \frac{1}{4} & \frac{5}{8} & 0 \\ 0 & \frac{1}{8} & \frac{c}{8} \end{pmatrix}$$

The eigenvalue of A is computed as

$$\begin{vmatrix} (\frac{1}{4} - v) & \frac{1}{4} & 0 \\ \frac{1}{4} & (\frac{5}{8} - v) & 0 \\ 0 & \frac{1}{8} & (\frac{c}{8} - v) \end{vmatrix} = 0$$

which gives the solutions

$$v_1 = \frac{3}{4} \quad v_2 = \frac{1}{8} \quad v_3 = \frac{c}{8}$$

If $c \leq 6$, then we get the same standard system as in Chapter 3. If $c > 6$, i.e. if the productivity in the C-sector is lower than the productivity in the whole subsystem which produces the basic products (A and B sectors together), then the standard system breaks down in the sense that we face the possibility of getting negative prices.

If, for example, $c = 7$, then we get the following weights in the standard system

$$q_1 = \frac{50}{72} \qquad q_2 = \frac{125}{72} \qquad q_3 = \frac{150}{72}$$

and $R = 1/8$.

If $c < 6$ then $q_3 = 0$ and $R = 1/3$.

In his communication with Newman (see Bharadwaj, 1970), Sraffa says that the "immense majority of non-basics are not used in production, not even in their own production: so they do not even form individual systems. Some (mainly animals and plants) are used each in its own reproduction and form individual systems. A few may be linked with one or two others, because of mixing, or cross-breeding, or if the length of gestation brings out the egg-hen dicotomy. And that is all. The third class, which is the least numerous and may be worth mentioning for the sake of completeness, is the source of all trouble."

APPENDIX B

The purpose of this appendix is to study the effects of changes in the income distribution on the prices in a Sraffa-system.

We assume the following system (only basic products).

$$\begin{cases} (A_a p_a + B_a p_b + \dots + N_a p_n)(1 + r) + L_a w = Ap_a \\ \vdots \\ (A_n p_a + B_n p_b + \dots + N_n p_n)(1 + r) + L_n w = Np_n \end{cases} \tag{7B.1}$$

This system can be transformed to dated quantities of labour (cf. Chapter 3). For a given product A we get

$$Ap_a = (1 - \frac{r}{R}) \left[L_a + L_{a1}(1+r) + \dots + L_{an}(1+r)^n \dots\right] \tag{7B.2}$$

Derivating (7B.2) with respect to r gives

$$A \cdot \frac{dp_a}{dr} = \frac{1}{R} \{- L_a + L_{a1} \left[(R-r) - (r+1)\right] + L_{a2}(1+r) \left[(2R-2r) - (r+1)\right] + \dots + L_{an}(1+r)^{n-1} \left[(nR-nr) - (r+1)\right] + \dots \tag{7B.3}$$

Using (7B.3) we make the following statements.

Statement 1: If $\frac{dp_a}{dr} > 0$ when r is large ($r < R$), then it is true that $\frac{dp_a}{dr} > 0$ for every r

proof: we assume that $\frac{dp_a}{dr} > 0$ when $r \rightarrow R$. This means that the paranthesis in (7B.3) is positive. If r decreases, then the number of positive terms in (7B.3) increases.

Statement 2: If $\frac{dp_a}{dr} < 0$ when r is very small, then it is true that $\frac{dp_a}{dr} < 0$ for every r

proof: we assume that $\frac{dp_a}{dr} < 0$ when $r \rightarrow 0$. This means that the paranthesis in (7B.3) is negative. If r increases, then the number of negative terms increases.

<u>Statement 3</u>: $\frac{dp_a}{dr}$ can change sign only once.

<u>proof</u>: the expression (nR-nr) - (r+1) can, for a given n, change sign once at most.

CHAPTER 8 WAGE EARNERS' FUNDS

In this chapter we shall discuss wage earners' funds using the method of analysis which we described in chapter 5. The end of chapter 5 is of specific importance here because it shows how financial markets can be included in post-Keynesian analysis.

Wage earners' funds are of great interest now since many countries have suggested various ways for constructing such funds . In Sweden the labor unions have become more interested in using wage earners funds to equalize ownership in the society and as an addition to other methods for increasing the power of labor in the firms.

In Sweden the issue is of special interest since a radical change in saving is taking place. During a long period of time there has been a gradual decline in corporate savings and the saving has instead been done by a public pension fund (see table 1). The country has, however, now reached a point at which the fund - on the whole - does not add any more net saving. If saving (and thereby the increase of wealth) does not take place in the public pension fund and if the savings of the household sector is relatively constant see table 1 the corporate saving is likely to increase. From the point of view of the distribution of wealth it is thus very important how the new saving is done. One possibility to distribute wealth and ownership more equal is to use wage earners' funds.

Year	Government	Pension fund	Financial corporations	Non-financial corporations	Households	Total mill.cr
1950-54	24.8	0.0	8.1	39.2	25.2	4.918
1955-58	32.3	2.8	6.3	23.1	38.5	6.743
1959-62	37.1	7.0	6.6	17.1	25.4	10.520
1963-66	37.4	18.1	3.7	10.3	23.5	16.875
1967-70	37.5	28.0	7.1	9.7	13.4	21.937
1971-73	30.5	33.2	9.7	11.0	26.0	28.377

Table 8.1 Percentage distribution of net saving

8.1 Some forms of wage earners' funds

We shall here briefly describe two different constructions of wage earners' funds which have been suggested in Germany and Denmark which are based on payments to the fund from the profit and total wages respectively.

a. The German system

According to a German suggestion which has been much discussed corporations must pay a certain fraction of their profit (which must be carefully defined) to funds owned by the employees. The percentage of the profit which shall be paid to the funds varies and start when the profit amounts to DM 400.000. When the profit gets larger the percentage goes up and is 10% when the profit is DM 1.000.000. (This means that approximately 12% of all joint stock companies in Germany must pay to the funds. The principal rule is that payment must be made in the form of shares.[1])

The shares owned by the employees will be kept in many funds. A wageearner can select in which fund he wishes to hold his shares. The funds invest the money for the employees and functions from this point of view like a mutual fund.

According to the original German suggestion the corporations will be allowed to deduct taxes, salary to the management,and dividends before payments to the funds are made.

b. The Danish Suggestion

All employees shall according to this proposition be owners of the firms independently of were they work and they shall receive the same number of shares independently of what their salary is.

When the system starts, all corporations shall pay 1% of all salaries to the funds, then the percentage goes up by 1/2% unit every years until the funds receives a permanent level of 5% of the total salary.

The employees obtain shares in the fund which holds their shares. They cannot sell their shares until after five years (except if

---

1) If a company prefers to pay cash rather than using shares the percentage goes up from 10 to 15%.

accidents or deseases would make this necessary).

The "Swedish system" is similar to the German in the sense that the payments to the funds are determined as a percentage of the profit (20%). The wage earners will however not own any individual shares in the fund.

8.2 Analysis

To continue the analysis we shall use the following notiations

- K = capital
- b = fraction of profit invested in the funds
- ρ = profit as a fraction of capital
- g = rate of growth
- i = fraction of growth financed by emission of new shares
- s_k = corporate saving (fraction of profit)
- s_w = wage earners propensity to save
- W = wages

Before analyzing the effects of a wage earners fund it might be of some interest to investigate, if the wage earners in Sweden automatically will take over capital according to what we discussed in chapter 5 section 5.5.

As was shown in chapter 5 an important problem is to investigate if the inequality

$$s_w \frac{Y}{K} < g_n \tag{8.1}$$

holds or not.

If it holds as an equality, i.a.

$$s_w \frac{Y}{K} = g_n \tag{8.2}$$

then the wage earners will take over all capital in society. Whether or not this is the case is an empirical problem. According to Edgren et al p. 132 ff the saving undertaken by households was 6% of disposible income during the period 1960-65. Included in this is savings in private homes which should not be included in s_w. Only saving which is available for investment in productive assets is considered.

$g_n \frac{K}{Y}$ was 17% during the same period. Later figures indicate that s_w should be slightly less than 6%. The conclusion is that the inequality (8.1) holds with a wide margin, i.a. the wage earners will not via their savings take over the capital in society.

One can however raise s_w by various methods and this might lead to a more satisfactory distribution of wealth in society.

In order to compute the share of capital in society owned by the wage earners in a stationary state we shall use the discussions in chapter 5 section 5.5

Let us assume that the profit can be divided in one part P_K which goes to the owners of capital and another part P_w which goes to labor. Total profit in society is the sum of P_K and P_w

If we denote the wage earners' propensity to save by s_r and that of the capitalists' by s_K we obtain total saving in society S, where we denote total wages by L.

$$S = s_w(L + P_w) + s_K P_K \tag{8.3}$$

and since total income Y can be written

$$Y = P_w + P_K + w$$

we can write (8.3) in the following way

$$S = s_w Y + (s_K - s_w)P_K \tag{8.4}$$

Since saving must equal investment i.e. I = S we can write (8.4) in the following way

$$\frac{P_K}{Y} = \frac{1}{s_K - s_w} \frac{I}{Y} - \frac{s_w}{s_K - s_w} \tag{8.5}$$

We know that in the long run the fraction of capital owned by labor equals their share of total savings i.a. we have that

$$\frac{K_w}{K} = \frac{S_w}{S} = \frac{s_w(Y - P_K)}{I} \tag{8.6}$$

where K is total capital, S total saving $S_w \{= s_w(L+P\ )\}$ is the workers' total saving and K_w is capital owned by the workers.

Using (8.5) we can write (8.6) in the following way

$$\frac{K_w}{K} = \frac{s_w s_K}{s_K - s_w}\,\frac{Y}{I} - \frac{s_w}{s_K - s_w} \qquad (8.7)$$

This can be written

$$\frac{K_w}{K} = \frac{s_w(s_K \frac{Y}{I} - 1)}{s_K - s_w} \qquad (8.8)$$

In figure 8.1 we have illustrated (8.8). We have set $\frac{I}{Y}$ equal to 0.17 and s_K to taken equal to 0.7. These figures seem to approximate the conditions in Sweden.

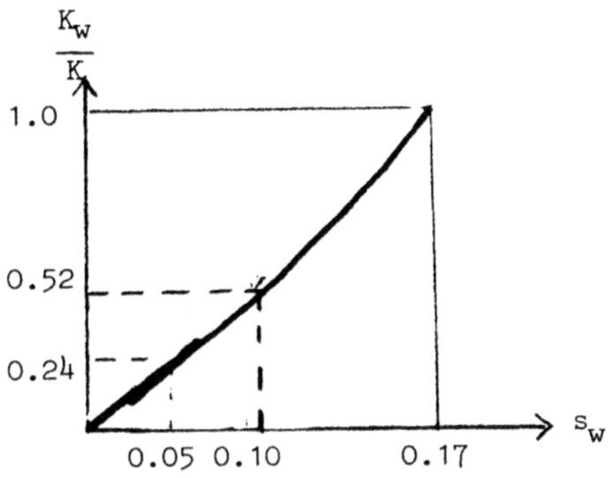

Figure 8.1

If one thinks that it is reasonable that the wage earners obtain approximately half of capital than we see from the figure that a propensity to save s_w = 0.1 gives $\frac{K_w}{K}$ = 0.52. We also see that s_w = 0.05 gives $\frac{K_w}{K}$ = 0.24 which might be sufficient to control many firms. However, in order for the employees to control the firms at this lower fraction of total capital they have to organize their ownership collectively in some way.

8.3 The valuation ratio

In addition to the shares obtained free as a given fraction of profits, it is usually stipulated that the fund shall buy newly emitted shares according to some rule. The issuing of new shares will affect the value of existing shares which is usually described by the valuation ratio.

Before discussing the determination of the valuation ratio we shall define the additional notations used.

N = number of shares

r = rate of return used for discounting future dividends

x = fraction of capital owned by the fund

y = fraction of capital owned by capitalists

s_K = corporate saving

The valuation ratio is defined as the ratio of the value of the shares and capital invested in firm

$$v = \frac{p \cdot N}{K} \tag{8.9}$$

The value of one share is the present value of all future dividends alotted to that share. If we begin by considering a situation where a wage earners' fund every year receives a fraction $b\rho$ of capital in the form of new shares, it follows that the value of one share is

$$(1-s_K)\rho \int_0^\infty \frac{K_o e^{s_K \rho t}}{N_o e^{b\rho t}} \cdot e^{-rt} dt \tag{8.10}$$

which gives the following value of every share

$$p = \frac{(1-s_K)\rho K_o}{(r+b\rho-s_K\rho)N_o} \tag{8.11}$$

It follows that the valuation ratio is

$$v = \frac{(1-s_K)\rho}{r+b\rho-s_K\rho} \tag{8.12}$$

The effect on the valuation ratio of new shares emitted depends on the price of the new shares. If the new shares have a price which equals the value of their future dividends they will not affect the valuation ratio.

8.4 The share of capital owned by the fund

In this section we shall determine the fraction of total capital in society owned by the fund. We shall assume that the fund receives a certain fraction every year of total profit as discussed above and that it reinvests part of the dividend it receives.

An important assumption is that the fund can buy all the shares which it desires at a price determined by the valuation ration. What the fund does not buy is bought by the capitalists. This must be the case since if they would not buy the remaining shares supplied, prices would drop below what is consistent with the valuation ratio and it is profitable to start buying.

Situation a

We begin the analysis by examining the rule proposed by Meidner 1975. He states that the fund shall buy a fraction of newly emitted shares equal to its fraction of total capital. Using the notations defined above, we obtain the following equation describing how the share of capital owned by the fund develops over time.

$$\frac{d(xK)}{dt} = xigK + b\rho K + x(s_K\rho - b\rho)K \qquad (8.13)$$

$$x(0) = 0$$

Equation (8.13) states that the capital owned by the fund grows by

a) what it buys of new shares
b) what it obtains free
c) its share of the growth of capital (after dividend and donation to the fund are deducted).

We know that capital is growing according to $K(t) = K_o e^{gt}$. Therefore we can write (8.13) as follows

$$\frac{dx}{dt} = x(ig + s_K\rho - b\rho - g) + b\rho \qquad (8.14)$$

$$x(0) = 0$$

In order for saving to be equal to investment, we must have that

$$igK + s_K\rho K = gK \qquad (8.15)$$

Using (8.15) we can write (8.14) as follows

$$\frac{dx}{dt} = - xb\rho + b\rho \tag{8.16}$$

and we obtain the solution

$$x(t) = 1 - e^{-b\rho t} \tag{8.17}$$

It follows directly that the share of capital owned by the capitalists y(t) will develop over time as

$$y(t) = e^{-b\rho t} \tag{8.18}$$

Our conclusion is that if the wage earners' fund can obtain the same rate of growth of its capital as the capitalists can and if they follow the rule for purchasing new shares suggested by Meidner (1975) then they will end up by owning all capital in society. It also follows that the fund will - on the average - own a majority in corporations after a time t^* determined by

$$1 - e^{-b\rho t^*} = e^{-b\rho t^*}$$

$$b\rho t^* = \log 2$$

$$t^* \simeq 23 \text{ years (using } \rho \simeq 0.15 \text{ and } b \simeq 0.2)$$

In discussions about the desirability of increasing labor's ownership of capital it is often argued (including Meidner's discussion) that workers should own a "fair" share of capital. An early example is Owen (1969) who states

> "That which can create wealth is of course worth the wealth which it creates ... Of this new wealth, so created, the labourer who produces it is justly entitled to his fair proportion." (1969, p. 222.)

It is clear from our results above that Meidner's rule will lead to a situation where labor owns all capital. If a smaller fraction is fair the rule must be formulated differently.

There clearly are many possibilities to do this but if we week one that does not change over time, it seems natural to alter the fraction of newly emitted shares bought by the fund.

Situation b

Assume that the fund saves a fraction s_F of the dividend that it receives. Equation (8.13) now takes the form

$$\frac{dxK}{dt} = s_F(1 - s_K)\rho Kx + b\rho K + x(s_K\rho - b\rho)K \tag{8.19}$$

Using (8.15) and solving (8.19) we obtain

$$x(t) = \frac{b\rho}{ig + \rho\{b - s_F(1-s_K)\}}\{1 - e^{-(ig+\rho\{b-s_F(1-s_K)\})t}\} \tag{8.20}$$

The parameters must fulfill $x < 1$ for all t. g can be eliminated using (8.15).

It directly follows that the share owned by capitalists is determined by

$$y(t) = \frac{b\rho}{ig+\rho\{b-s_F(1-s_K)\}} e^{-\{ig+\rho(b-s_F(1-s_K))\}t} + \frac{ig-s_F(1-s_K)\rho}{ig+\rho\{b-s_F(1-s_K)\}} \tag{8.21}$$

The valuation ratio is taken as given here and we have assumed that the fund has stated a rule by which it wants to buy shares. The rest of the shares are then bought by the capitalists. This will mean that their propensity to save varies over time. Considering our example above we have assumed that they buy the emitted shares not bought by the fund,i.e. they buy every year

$$igK - s_F(1-s_K)\rho Kx$$

where x is determined by (8.20). They receive a dividend equal to $(1 - s_K)(1 - x) K$. Therefore their propensity to save is taken to be

$$\frac{i_g - s_F(1-s_K)\rho x}{(1-x)\rho(1-s_K)}$$

In the long run labor will own a "fair" share

$$x = \frac{b\rho}{ig+\rho\{b-s_F(1-s_K)\}} \tag{8.22}$$

If b, s_F, and s_K are fixed and one desires a certain value of x, then we see directly from (8.22) that the rate of profit will be determined since g can be eliminated using (8.15).

8.5 The capitalists' propensity to save

Our procedure has been to use a fixed valuation ratio determined by the present value of future dividends assuming that there exists an exogenous rate of time preference r. We have assumed that the fund decides what part of new shares that it wants to buy and that the "old" capitalists buy the remaining ones.

Situation a is perhaps most interesting since it has been suggested by Meidner (1975) and since the right to buy new shares is proportional to old shares. It does, however, follow that we assume that the capitalists have a propensity to save $\bar{s}_K$ equal to (use 8.15)

$$\bar{s}_K = \frac{s_K \rho K + igK}{\rho K} = \frac{g}{\rho} \tag{8.23}$$

If $\bar{s}_K < \frac{g}{\rho}$ then one has to introduce other agents into the system who buy the shares not bought by the fund and the capitalists. If $\bar{s}_K > \frac{g}{\rho}$ then we must assume that there exist investments outside of the corporate sector studied.

In situation b the capitalists buy the new shares not bought by the fund, i.e. they buy $igK - s_F(1-s)\rho Kx$. Thus their propensity to save becomes

$$\bar{s}_K = \frac{ig + xs\rho - s_F(1-s_K)\rho x}{(1-x)\rho} \tag{8.24}$$

which is a function of time.

Equation (8.24) naturally is a very strong assumption to make and the system studied must be extended to include other sectors of the economy as discussed above.

We have here, as mentioned above, used a different way to determine the valuation ratio than in section 5.6. The valuation ratio has here been taken to be determined by long run dividend policy and not by short run market conditions.

If we instead should determine v by equation (5.30) where we for the sake of simplicity assume that $s_w = 0$ we obtain

$$0 = xc(vgK - igK) + xigK$$

e.g. the "old" shareholders own a fraction x of capital and buy a corresponding fraction of newly emitted shares

We obtain that v is determined by the same equation as if no investment in wage earners funds was made. (See 5.32 with $s_w=0$.)

$$v = \frac{i(c-1)}{c}$$

We have here made the special assumption that c is constant i.e. that the "old" earners of capital always consume (or save if c is negative) the same fraction of capital growth <u>independent of</u> what fraction of total capital that they own.

8.6 Wage earners' funds and the stockmarket

By creating a fund which is collectively owned and which every year obtains a fraction of the profit in firms and save a large part of what it obtains as owner of capital the wage earners get a very much increased propensity to save. Can the saving now become so large (compare the discussion in section 8.c.) that the wage earners take over all capital?

There are a few aspects which makes this doubtful. Many suggestions leave out small firms.[1] It is difficult to find data which show that small firms grow quicker than large ones even if this has been said to be the case, see Kaldor (1966). But much of the dynamics in the economy naturally comes from new enterprises where fortunate capitalists can enjoy a higher rate of growth than what the wage earners fund can earn on its shares in older enterprises.

As was mentioned above the funds often will use their dividend income to buy new shares emitted by the firms.

---

1) It is likely that any such limit will influence the size distributions of firms in an economy. When a firm approaches the limit for paying to a wage earners' fund it might stop growing and instead the old owners start a new enterprise.

Equation (8.15) can be written

$$\rho = \frac{g(1-i)}{s_K} \qquad (8.25)$$

From (8.25) we see the importance of i. We have taken g to be fixed. The parameters i and s_K are controlled by the firms. We have shown above (in section e.) some important aspects of the ways in which new shares are bought. We shall here only add some problems connected with employment.

A wage earners' fund is likely to have effects on employment because the enterprises could influence the growth of the fund by not employing people. We have in the analysis above assumed full employment. But to have full employment it is probably necessary that the rate of profit exceeds a certain minimum value.

$$\rho \geq \rho_{min} \qquad (8.26)$$

where ρ_{min} is a value which the rate of profit must exceed in order for enterprises to consider investments worthwhile. From (8.25) we see that ρ_{min} puts a limit on the parameter i.

If ρ_{min} will change is hard to say but it seems clear that many countries experience a fear of capital moving out of the country as a result of introducing wage earners funds something which would effect ρ_{min} and cause unemployment.

8.7 Conclusions

We have separated the population in wage earners and earners of capital even if the dividing line between them is not sharp. Many persons are both wage earners and owners of capital. Wealth is however distributed in a very uneven way and it might be reasonable to use the approximation that a small fraction of society owns a large fraction of capital and obtains all income from capital.

The Swedish debate about wage earners' funds, which is related to similar discussions in other countries, starts from an assumption that the profit is a residual which can be distributed in any wayas long as one makes sure that total saving in society is so large that sufficient growth and economic stability is maintained.

The wage earners state what share of capital they think is reasonable to own and they shape a system for collecting parts of corporate profits and construct rules for individual use of the "labor capital" such that the necessary amount of saving is guaranteed. The wage earners are assumed to have sufficient political power to introduce the new system.

We have shown here that important aspects of such a new system can fruitfully be analyzed by using some methods described in Part 1 specifically in Chapter 5.

CHAPTER 9 TECHNOLOGICAL CHANGE

9.1 Introduction

In growth economics problems are discussed on a highly aggregated level. The emphasis is on changes in output, level of employment, and physical capital in a growing economy. The distribution of income between labour and capital is also of interest.

In Chapters 5 and 8 we studied the distribution of income in a growing economy and different methods for redistributing the ownership of the production means. In this chapter we will continue studying economic growth. However, the emphasis will be on technological change.

As was pointed out in Chapter 2, Ricardo showed a limited interest in technological change, at least in comparison with Smith and Marx. We will, however, in this chapter investigate how Ricardo's concepts and neo-Ricardian concepts and methods can be used in order to critisize the neo-classical treatment of technological change (Section 9.2). Furthermore, we will present a growth model which is based on Ricardian concepts in its treatment of technological change (Section 9.3).

When we try to interpret the concept "technological change" we face problems which are similar to those we faced in trying to interpret the concept "inflation". Firstly, the concept "technological change" can be used in order to refer to <u>effects</u> of changes in the technology. In that case one is interested in the role played by technical progress in economic growth. Secondly, the concept can refer to <u>changes</u> in technology itself, defining technology as "useful knowledge pertaining to the art of production" (Kennedy and Thirlwall, 1972).

Effects of changes in technology, the first interpretation, are very often highly valued because they appear to lead to improved standards of living. During the last years the negative effects have been more apparent, e.g. pollution and problems for people when their knowledge and experience have become obsolete.

Referring to the first interpretation of technological change, we can say that technological change can take at least three forms:

(a) new products
(b) new production processes
(c) new administration methods

To measure technological change in all of its forms is not possible. For example, how does one measure the level of knowledge or growth in knowledge? Instead of measuring technological change directly one tries to measure it indirectly as a residual. In other words, the increase in output which cannot be explained by the increase in input is referred to as the effect of technological change. This measure of technological change cannot tell us, of course, if the change is positive or negative in the sense discussed above.

When using time-serie data in estimating technological change, one faces a methodological problem, namely how to separate returns to scale from technological change. In all empirical work, based on neo-classical theory, it is assumed that it is possible to distinguish between capital accumulation and technological change. This means that one has assumed that it is possible to distinguish between movements along a production function (accumulation) and shifts in a production function (technological change).

The purpose with Section 9.2 is to show that the neo-classical measure of technological change is wrongly defined at least if one wishes to use it to forecast price changes (which seems to be a modest claim). In Section 9.2 we utilize the work of Rymes (1971) which can be said to contain the dynamic implications of Sraffa's criticism of the neo-classical theory.

Above we have only discussed the effects of technological change but it is obvious that this change must be based on (cf. interpretation 2 above):

(1) research and development activity
(2) education
(3) means for implementing new technology

In the growth models, which can be found in the literature, a common trait concerning the majority of them seems to be that technological change is treated as an exogenous variable. The technological change is a function of time only. If all investments are stopped there will still be technological progress. It means that technological knowledge is falling like manna from heaven (see Hahn and Matthews, 1970, p.377). A refinement of the models mentioned above are the so-called vintage models. In these models the manna falls only on the machines which have been manufactured during the last period.

In a second class of models, technological change is considered as an endogenous variable, i.e. the rate of technological change depends on other economic variables. Arrow (1962) relates technological change to accumulated gross investments. It means that all investments contain an element of investment in research and development. Arrow's model, like the one we are going to discuss in Section 9.3 (Kaldor and Mirrlees, 1962), includes an element of learning. New products have varying efficiency depending on when they are produced. We get production equipments of different vintages and where the most modern equipment also is the most efficient.

9.2 Measurement of technological change: a critique

The most common indicator of technological change is some kind of measure of factor productivity. Les us assume a Cobb-Douglas function

$$Q = FL^aK^b \qquad a + b = 1 \qquad (9.1)$$

where (9.1) is true of a one-sector economy or an aggregated version of a multisector economy (in that case the criticism discussed in Chapter 4 is still valid).

Q = output
F = factor productivity
L = labor
K = capital

(9.1) can be written as

$$\log Q = \log F + a \log L + b \log K \qquad (9.2)$$

Derivating (9.2) with respect to time, t, gives

$$\frac{dQ}{dt} \cdot \frac{1}{Q} = \frac{dF}{dt} \cdot \frac{1}{F} + a\frac{dL}{dt} \cdot \frac{1}{L} + b\frac{dK}{dt} \cdot \frac{1}{K} \qquad (9.3)$$

or, in an alternative notation

$$\frac{\dot{Q}}{Q} = \frac{\dot{F}}{F} + a\frac{\dot{L}}{L} + b\frac{\dot{K}}{K} \qquad (9.4)$$

A common trait in growth economics is the study of rate of change, and that is what (9.4) expresses. (9.4) can be written as

$$\frac{\dot{F}}{F} = \frac{\dot{Q}}{Q} - a\frac{\dot{L}}{L} - b\frac{\dot{K}}{K} \qquad (9.5)$$

The rate of change in factor productivity, $\dot{F}/F$, is a measure of technological change. In (9.5) technological change is a residual, which is unaccounted for when we have considered changes in labour and capital. It should also be mentioned that in (9.5) capital is treated as an input similar to labour. As will be shown later, this causes trouble for the neo-classical measure of technological change. Furthermore, the change in total factor productivity is interpreted as a shift in the neo-classical production function. The change in K is interpreted as a movement along the production function.

The most modern approaches of measuring total factor productivity, e.g. Jorgensen and Griliches, take the social accounts as a point of departure. There we have the following identity in an one-sector economy

$$PQ \equiv WL + PRK + DPK \qquad (9.6)$$

where

P = the nominal price level

W = wage per unit of labour input

R = the net rate of return to the stock of commodity capital

D = the rate of depreciation

L = the flow of labour input

K = the stock of commodity capital

Q = output of the commodity

Derivating with respect to t gives

$$\dot{P}Q + P\dot{Q} \equiv \dot{W}L + W\dot{L} + \dot{P}RK + P\dot{R}K + PR\dot{K} + \dot{D}PK + D\dot{P}K + DP\dot{K} \quad (9.7)$$

which can be written as

$$\frac{\dot{P}}{P}PQ + \frac{\dot{Q}}{Q}PQ \equiv \frac{\dot{W}}{W}LW + \frac{\dot{L}}{L}WL + \frac{\dot{P}}{P}PRK + \frac{\dot{R}}{R}PRK + \frac{\dot{K}}{K}PRK + \frac{\dot{D}}{D}DPK + $$

$$+ \frac{\dot{P}}{P}DPK + \frac{\dot{K}}{K}DPK \quad (9.8)$$

If we use lower-case letters to represent rate of change, e.g. $\frac{\dot{P}}{P} = p$, $\frac{\dot{Q}}{Q} = q$ then we get

$$p + q \equiv \alpha(w+l) + \beta(p+r+k) + \gamma(d+p+k) \quad (9.9)$$

where

$$\alpha = \frac{WL}{PQ}, \; \beta = \frac{PRK}{PQ}, \; \gamma = \frac{DPK}{PQ}$$

(9.9) can be expressed as

$$q - [\alpha l + \beta k + \gamma(d+k)] \equiv [\alpha w + \beta(r+p) + \gamma p] - p \equiv t \quad (9.10)$$

where t is the residual, the rate of technological change. (9.9) and (9.10) are written in what is called Divisia index number form. Rymes (1971, p.56) gives the following interpretation of (9.10): "The growth rate of total output is represented by q and the growth rates of the appropriately weighted quantities of the inputs by $[\alpha l + \beta k + \gamma(d+k)]$. Hence the difference must represent, in some sense, the growth rate of the efficiency or technology of the economic system. The appropriately weighted growth rate of prices of the inputs is $[\alpha w + \beta(r+p) + \gamma p]$ and the growth rate of price of output is p. Again the difference must be equal to the growth rate of the efficiency or technology of the economic system."

The left-hand side of (9.10) is the measure of technological change used by Jorgenson and Griliches, correctly computed and expressed as rate of change. According to Rymes all measures of technological change are based on some version of (9.10).

The left-hand side of (9.10) can be used to compute what is called "the sources of growth". We get

$$1 \equiv \frac{\alpha l}{q} + \frac{\beta k}{q} + \frac{\gamma(d+k)}{q} + \frac{t}{q} \qquad (9.11)$$

where for example $\frac{\alpha l}{q}$ is said to be labour's contribution to growth. (9.11) tells us nothing about the causes of growth, it is only an ex post accounting. If the first three terms in (9.11) do not add up to 100 %, then the rest is said to be due to technological change. It is of course more interesting to know to what extent a measure of technological change can be used to predict price changes. To that question we will return.

It should be mentioned that in using (9.10) we have to face the problem of measuring the value of capital. The reason is that we cannot compute the reutrn on capital if we do not know the prices of capital commodities.

The interpretation of K which is used above, when capital means capital commodities and where these capital commodities are inputs, either as flow or stock, is one of the definitions of capital which Rymes utilizes. According to Rymes this is the definition used in modern neo-classical theory (Walras-systems which do not need any assumptions about aggregated capital or "smooth" production functions). It is this interpretation of capital as a primary input that Rymes want to show is false, i.e.

that "the capital controversy" (discussed in Chapter 4) goes beyond the neo-classical distribution theory.

Rymes' second definition of capital as input is real capital, which is the capital concept used by Harrod and Robinson. This concept Rymes traces back to Ricardo who wrote:

> "... capital may increase without its value increasing, and even while its value is actually diminishing not only may an addition be made to the food and clothing of a country, but the addition may be made by the aid of machinery, without any increase, and even with an absolute diminution in the proportional quantity of labour required to produce them. The quantity of capital may increase, while neither the whole together, nor any part of it singly, will have a greater value than before, but may actually have less."
> (Ricardo, 1971, p.117)

If K is capital interpreted as capital commodity, then

$$J = \frac{K}{W/P} \qquad (9.12)$$

is capital interpreted as real capital. (9.12), which is Joan Robinson's capital concept, gives us the amount of capital measured in labour time. That capital concept makes sense if the capital commodity is seen as something that will assist labour in production. However, the measure is not independent of income distribution and prices. Another concept which is needed before we continue the analysis is steady-state, in this case dynamic steady-state. A steady-state concept in this context is "neutral technical progress". The purpose of such a concept is to indicate characteristics of technical or technological change that in some sense leave unchanged the balance between capital and labour and will permit a steady growth (cf. Hahn and Matthews, 1970, p.367). The two most common definitions of neutral technical change are

(a) Hicks-neutral change; implies that K/L is constant

(b) Harrod-neutral change; implies that the rate of profit is constant.

The purposes of the two definitions are not the same. Hicks wanted to explain the observed constancy in the distribution of income between labour and capital. He postulated a constant K/L and investigated the implications of that assumption. Harrod, however, was interested in developing a dynamic equilibrium theory and looked for a definition of neutrality which was suitable for that purpose. The majority of the growth models do not assume a Hicks-neutral progress, instead they are based upon Harrod's definition. However, for a Cobb-Douglas function, cf. (9.1), the two definitions coincide.

Above we discussed neutrality in one-sector models or aggregated models. The concept of neutral technical progress becomes more complicated in a model that has more than one commodity. If the whole economy in a multi-sector model gets a neutral progress is dependent not only of the neutrality in each sector but also of the variation in the rate of change between different sectors.

A Harrod-neutral invention (which gives a Harrod-neutral progress) can be interpreted to give an increase in efficiency of labour in the whole economy. This can be written as

$$Q = f(K, \alpha(t) \cdot L) \tag{9.13}$$

A Harrod-neutral progress has the same effect as an increase in the population (and it is also sometimes called "labour-augmenting"). We have no reason to assume that the technological change in the real world is neutral, and the model we present in the next section does not rely on a neutrality concept. The reason that Rymes assumed Harrod-neutral progress, constant returns to scale, perfect competition, and full employment is that he wants to show that even when these assumptions hold,the neo-classical analysis of capital accumulation and technological change is wrong. Neo-classical analysis is most comfortable when these assumptions hold. Below we present a somewhat simplified version of Rymes' analysis.

Let us assume a two-sector economy with one consumption commodity sector (the c-sector) and one capital commodity sector (the k-sector).

Consider the equilibrium accounts for this two-commodity and two-sector model

$$P_cC \equiv WL_c + RP_kK_c$$
$$P_kM \equiv WL_k + RP_kK_k \tag{9.14}$$

where

C = output of the consumption commodity
M = output of the capital commodity
K_c = number of units of the capital commodity used in the c-sector
K_k = number of units of the capital commodity used in the m-sector
P_c = price per unit of the consumption commodity
P_k = price per unit of the capital commodity
W = the wage per unit of labour
R = return on capital
L_c = labour units used in the c-sector
L_k = labour units used in the k-sector

We assume that the quality of the commodities produced does not change over time and we ignore depreciation. The purpose is to find out if it is possible to separate capital accumulation from technological change, where in this context we interpret technological change as improved production mehtods. Derivating with respect to time, we get (cf. (9.9))

$$\begin{cases} p_c + c \equiv \alpha_c(w+l_c) + \beta_c(r+p_k+k_c) \\ p_k + m \equiv \alpha_k(w+l_k) + \beta_k(r+p_k+k_k) \end{cases} \tag{9.15}$$

(9.15) can be written in Divisia index number form, and if we at the same time assume that L_c and L_k are constant (remember the purpose mentioned above), then we get

$$\begin{cases} c - \beta_ck_c \equiv [\alpha_cw + \beta_c(r+p_k)] - p_c \equiv t_{CNC} \\ m - \beta_kk_k \equiv [\alpha_kw + \beta_k(r+p_k)] - p_k \equiv t_{KNC} \end{cases} \tag{9.16}$$

where t_{CNC} expresses the technological change in the c-sector according to the neo-classical measure of technological change. We know that $\alpha_c + \beta_c = 1$ and $\alpha_k + \beta_k = 1$, then the left-hand side of (9.16) can be written as

$$\begin{cases} c - k_c + \alpha_c k_c \equiv t_{CNC} \\ m - k_k + \alpha_k k_k \equiv t_{KNC} \end{cases} \tag{9.17}$$

We have assumed Harrod-neutrality which means that

$$\begin{aligned} c - k_c &= 0 \\ m - k_k &= 0 \end{aligned} \tag{9.18}$$

We also introduce another steady-state assumption which is that the rate of change in output per labour unit is the same in the two sectors, i.e.

$$m = c \tag{9.19}$$

Now we get the solution of (9.17)

$$\begin{aligned} t_{CNC} &\equiv \alpha_c c \\ t_{KNC} &\equiv \alpha_k m \end{aligned} \tag{9.20}$$

Even if the rate of change per unit of labour is the same in both sectors, there is no reason to assume that $\alpha_c = \alpha_k$. In other words, we are just lucky if the neo-classical measure of technological change is the same in the two sectors. Such a measure cannot be used without adjustments if the purpose is to predict changes in relative prices.

An alternative measure of technological change is Harrod's measure, which is related to Robinson's concept "real capital". Rymes formulates the difference between the two measures of technological change as follows:

> "The neo-classical measure asks: Given the rate of technological progress, what is the rate of change of the primary and intermediate inputs required to maintain the rate of growth of output? The Harrod measure asks, <u>in addition</u>: Given the rate of technological progress, what is the rate of change of the primary and intermediate inputs required to maintain the rate of growth of the <u>intermediate inputs</u>?"
> (Rymes, 1971, p.98)

If we for a moment return to the one commodity model we studied in the beginning of this section, and if we add the assumptions we have introduced in the two-commodity model, the (9.10) can be written as

$$q - \beta k \equiv t_{NC} \tag{9.21}$$

Now, if we use the Harrod measure, then we get

$$q - \beta(k-t_R) \equiv t_R \tag{9.22}$$

where t_R is Harrod's measure of technological change. Behind the expression (9.22) we have a function of the following type (given some steady-state assumptions)

$$Q = Q(L,\ K/T_R;\ T_R) \tag{9.23}$$

K/T_R is capital defined in terms of some original factors (cf. Joan Robinson's "real capital"). Looking at (9.22), we also see how important it is to make a distinction between capital as input and capital as output. In (9.22) q and k refer to some commodity. Even if the commodity looks the same over time it is changing in original factors.

Harrod's measure of technological change, applied to our two-sector model, gives (see the left-hand side of (9.16))

$$\begin{cases} c - \left[\beta_c(k_c - t_{KR})\right] \equiv t_{CR} \\ m - \left[\beta_k(k_k - t_{KR})\right] \equiv t_{KR} \end{cases} \tag{9.24}$$

In (9.24) the interdependencies in the production process have been observed in a way related to Sraffa's system. Capital input in the c sector has been adjusted for the technological change in the k-sector.

(9.24) can be written as

$$\begin{cases} t_{CR} - \beta_c t_{KR} \equiv c - \left[\beta_c k_c\right] \equiv t_{CNC} \\ (1-\beta_k)t_{KR} \equiv m - \left[\beta_k k_k\right] \equiv t_{KNC} \end{cases} \tag{9.25}$$

why we get

$$\begin{cases} t_{CR} \equiv t_{CNC} + \dfrac{\beta_c}{1 - \beta_k} t_{KNC} \\ t_{KR} \equiv \dfrac{t_{KNC}}{1 - \beta_k} \end{cases} \qquad (9.26)$$

Using (9.20) we can write (9.26) as

$$\begin{cases} t_{CR} \equiv \alpha_c + \beta_c m \\ t_{KR} \equiv m \end{cases} \qquad (9.27)$$

The first identity in (9.27) can be written as $\alpha_c c + \beta_c m = \alpha_c(c-m) + m$, because $\alpha_c = 1 - \beta_c$.

If we recall the steady-state assumption which said that the rate of change in output per unit of labour is the same in both sectors, which means that $c = m$, then we get

$$\begin{cases} t_{CR} \equiv c \\ t_{KR} \equiv c \end{cases} \qquad (9.28)$$

In other words

$$t_{CR} = t_{KR} \qquad (9.29)$$

which should be compared with (9.20).

Harrod's measure of technological change gives the same rate of technological change in the two sectors (under the given assumptions). Therefore, it should be possible to predict that the relative prices are constant. That it is possible will be shown below.

However, let us first assume that the rate of technological change in the two sectors are different, ie. $t_{CR} \neq t_{KR}$, but that the progress in each of the two sectors is Harrod-neutral (we still asume that $l_c = l_j = 0$). Then it is possible to compute $p_k - p_c$ by utilizing (9.15). We get

$$p_k - p_c \equiv \alpha_c(c-m) \tag{9.30}$$

We can also see that if $t_{CR} = t_{KR}$, i.e. $c = m$, then we can predict that the relative prices are constant.

If we compute $t_{CR} - t_{KR}$ utilizing (9.27) we get

$$t_{CR} - t_{KR} \equiv \alpha_c c + \beta_c m - m$$

or

$$t_{CR} - t_{KR} \equiv \alpha_c(c-m) \tag{9.31}$$

Similar computations for the neo-classical measure of technological change gives, if we use (9.17)

$$t_{CNC} - t_{KNC} \equiv \alpha_c c - \alpha_k m \tag{9.32}$$

As we can see, the Harrod-measure (9.31) will exactly predict the relative price changes (9.30). There is no reason to suspect that the neo-classical measure (9.32) will predict even the direction of change in the relative prices. The situation in which (9.31) and (9.32) give the same result is when $\alpha_c = \alpha_k$. That is the situation when we assume a one-commodity economy.

Finally we study changes in relative prices using the right-hand side of (9.16). Rymes argues that if technological progress is to take on any meaning then, in a one-commodity model, the rate of growth of the input price should be greater than the rate of growth of the output price (Rymes, 1971, p.129). In other words, in a one-commodity model we have to make a distinction between input prices and output prices. According to Rymes, neo-classical theory has made a mistake in not paying attention to that distinction. However, Harrod's measure does pay attention to that distinction. In the two-sector model we get

$$\begin{cases} \alpha_c w + \beta_c(r+p_k+t_{KR}) - p_c \equiv t_{CR} \\ \alpha_k w + \beta_k(r+p_k+t_{KR}) - p_k \equiv t_{KR} \end{cases} \tag{9.33}$$

or

$$\begin{cases} \alpha_c(w-p_c) + \beta_c(r+p_k-p_c+t_{KR}) \equiv t_{CR} \\ \alpha_k(w-p_k) + \beta_k(r+t_{KR}) \equiv t_{KR} \end{cases} \tag{9.34}$$

Assuming Harrod-neutrality, the following solution is obtained

$$t_{CR} \equiv w - p_c$$

$$t_{KR} \equiv w - p_k$$

or

$$t_{CR} - t_{KR} \equiv p_k - p_c \tag{9.35}$$

In (9.35) we again find the fundamental relation between Harrod's measure of technological change and changes in relative prices.

Above we have given a simplified version of Rymes' analysis. He introduces more than two sectors, depreciation and quality changes, but the result of the analysis is the same. The neo-classical measure of technological change is incorrect. Rymes' conclusion is that much of the theoretical and empirical work which has been done in this area has to be written off as theoretically incorrect.

9.3 A growth model: Kaldor and Mirrlees

The model of Kaldor and Mirrlees (1962), which will be presented in this section, is an attempt to build a growth model in which we do not have to make a distinction between a shift in a production function and a movement along a production function. In this model, the firms want to cut their costs, and when they try to cut the costs they do not care whether they use neutral inventions or not. The model is a macro model based on verified behaviour of firms. It gives one aspect of the employment problem without explaining the level of employment.

We are going to use the following notations

n_t = the number of workers available to operate <u>new</u> equipment, i.e. equipment of vintage t. These workers cannot be taken from older equipment which is still profitable.

N_t = the working propulation at t

I_t = gross investments in fixed capital of vintage t

i_t = investment per operative on machines of vintage t

Y_t = gross national product at t

y_t = output per head

p_t = output per worker on latest machine. This output is constant for a given machine during its lifetime.

T = the lifetime of the equipment

w^*_τ = the expected rate of wages, which is a rising function of future time

r = the profit rate

We have the following identities

$$i_t \equiv \frac{I_t}{n_t} \qquad y_t \equiv \frac{Y_t}{N_t} \tag{9.36}$$

Kaldor and Mirrlees assume that the technological progress consists of two elements

(a) an exogenous flow of ideas

(b) the ideas are exploited in the investments

Assumption (a) means that the ideas are falling like manna from heaven whereas assumption (b) indicates that resources are needed in order to implement new technology. The relation between growth in output and investment is given by a technical progress function

$$\frac{\dot{p}_t}{p_t} = f\left(\frac{\dot{i}_t}{i_t}\right) \tag{9.37}$$

where

$f(0) > 0$

$f' > 0$

$f'' < 0$

$\dot{p} = dp/dt$, and $\dot{i} = di/dt$

In Figure 9.1 expression (9.37) is illustrated.

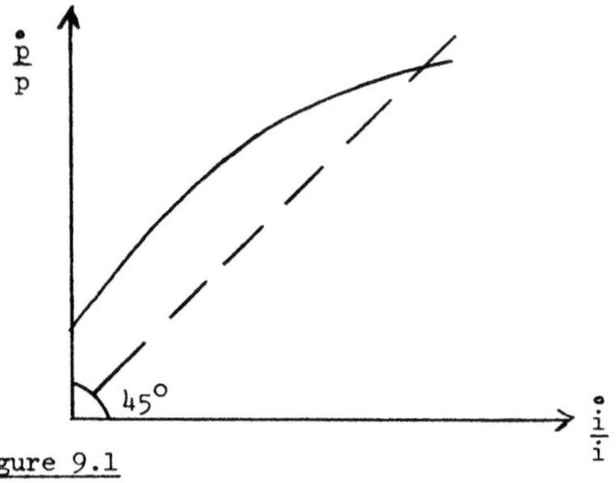

Figure 9.1

(9.37) is valid for equipment of the latest vintage.

In (9.37) we recognize Ricardo's interpretation of the law of diminishing return. Ricardo applied the law on land of different qualities. Kaldor and Mirrlees use the law to describe ideas of different quality (or different profitability). In Ricardo's theory, total land was given which together with the law of diminishing return and his wage theory converged towards the stationary state. In the model of Kaldor and Mirrlees the growth process is very dependent of the flow of ideas.

From Machlup (1962) we get some support concerning (9.37). He means that for a given level of knowledge (which can be represented by the above mentioned ideas) we meet diminishing returns of other inputs used (in our case investments). This gives us the conditions $f' > 0$ and $f'' < 0$. The condition $f(0) > 0$ means that output p is growing exponentially over time for a constant investment rate. An interpretation is that the number of problems which can be solved increases with the level of knowledge. It is also reasonable to expect that the cost of solving a given problem decreases with knowledge. Furthermore, Kaldor and Mirrlees assume that the firm invests if (here we simplify the analysis of Kaldor and Mirrlees)

$$i_t \leq \int_t^{t+T} e^{r(\tau-t)}(p_t-w_\tau^*)d\tau \qquad (9.38)$$

In other words, the firm demands that investment per worker gives a non-negative present value.

Another assumption is also introduced in the model, namely that the firm is also using a pay-back restriction. This assumption is formulated as

$$i_t \leq \int_t^{t+h} (p_t-w_\tau^*)d\tau \qquad (9.39)$$

which means that the cost of the fixed assets must be recovered within a certain period, in this case within h years.

(9.39) can be interpreted as a mean to notice technological change and risk. A fast technological progress affects the economic lifetime of production equipment. This, and the risk associated with technological change, can be noticed by h. If the investment is recovered then the firm has resources which can be used when exploiting new ideas (see Sellstedt, 1972, how restrictions like (9.38) and (9.39) can be used on the micro-level in the selection of research and development projects).

Kaldor and Mirrlees assume that (9.38) is satisfied whenever (9.39) is satisfied. This means that the model studies the growth process in an economy, where the firms' investment decisions are based on the pay-back principle.

The model is Keynesian in the sense that investment decisions are the primary growth mechanism. If we assume that all the savings come out of profits, then the share of profits in the gross national products is given by

$$\pi_t = \frac{1}{s} \cdot \frac{I_t}{Y_t} \tag{9.40}$$

where s = the proportion of profit that is saved.

We have the following relationship for the distribution of the labour force and for total output

$$N_t = \int_{t-T}^{t} n_\tau d\tau \tag{9.41}$$

$$Y_t = \int_{t-T}^{t} p_\tau n_\tau d\tau \tag{9.42}$$

The workers' share of total output is

$$Y_t(1-\pi_t) = N_t w_t \tag{9.43}$$

where w_t = the wage rate at t.

The equipment will be used only as long as the output is larger than or equal to the wages. This means that the profit for the equipment of the oldest vintage is zero. We get

$$p_{t-T} = w_t \tag{9.44}$$ [1]

(This expression can be compared with Ricardo's statement that rent on the worst part of the land utilized is zero.)

---

1) For a discussion of (9.44) see Nuti (1969) and Kaldor (1970).

Adding some assumptions about the population growth and the expected wage increase, the solution of the model gives dynamic stability only if

$$\frac{\dot{p}}{p} = \frac{\dot{i}}{i} = \gamma \text{ (constant)} \tag{9.45}$$

In other words, equilibrium is reached at the intersection point of the technical progress function, f, and the 45° line in Figure 9.1.

The equilibrium will in general be stable, but instability cannot be excluded. As Kaldor and Mirrlees points out "a downward drift of the technical progress function might allow the rate of growth of p to fall off, and remain below the rate of growth of w (which reflects the rate of growth of y over the recent past) sufficiently long until with falling investment, unemployment and stagnation set in." (1962, p.180)

Here we can see that it is almost possible to treat Ricardo's theory about the stationary state as a special case of the model of Kaldor and Mirrlees. In Ricardo's theory the quality of marginal land could only decrease when the area utilized increased. One interpretation of this is that the function f will drift downwards over time.

Conversely, an upward shift in the technical progress function (e.g. a result of a technical break-through) might lead to an inflationary situation. However, if the rate of growth at p does not differ to much from the rate of growth at w, then there will be a convergent movement until expression (9.45) is obtained.

Finally, we will touch upon some aspects of the employment problem. Derivating (9.41) with respect to time gives

$$n_t = \dot{N}_t + n_{t-T}\left(1 - \frac{dT}{dt}\right) \tag{9.46}$$

The labour force which is available for new equipment of vintage t consists of two components

(a) the growth in working population, $\dot{N}_t$

(b) the labour released by the retirement of obsolete equipment (T years old)

The model assumed full employment. However, (9.46) points out some potential problems. Will the firms treat all the individuals composing n_t as suitable for the new equipment? If the firms consider some individuals as not suitable and if we at the same time have an upward shift in the progress function, then the model indicates stagflation, i.e. inflation and unemployment at the same time.

9.4 Summary

In this chapter we have been dealing with technological change. We started with the concept "technological change". We found, utilizing Rymes' work, that the neo-classical concept is incorrect. In a dynamic economy it is not possible to make a distinction between shifts in a production function (technological change) and movements along a production function (accumulation). The analysis was based on neo-Ricardian methods. An alternative concept was presented. This concept was based on Ricardian theory.

Finally we presented a growth model which paid attention to technological change. This model was based on Ricardian concepts.

CHAPTER 10 THE EFFICIENCY OF DIFFERENT ECONOMIC SYSTEMS

Comparisons between economic systems of various kinds - capitalistic, socialistic,economies based upon labor managed firms have played an important role in the economic debate, see e.g. Bergson (1971) and Grossman (1967). This interest was specifically active in connection with the great depression around 1930. The decline of production and employment led to demands for a greater degree of central direction and planning of the economies. When economic systems were compared, difficulties arose as to the meaning of an economic system. (E.g. is it more important what the relations between enterprises look like or is it more important how the enterprises are internally organized?) When economic systems are compared this has been done along several dimensions, e.g.

(1) How ownership is distributed
(2) How large economic liberty is
(3) How the members of society are motivated
(4) How the economic units are coordinated and how resource allocation is organized
(5) Where different decisions are made in the economic system.

When different economic systems are evaluated, many objectives are involved, e.g.

(a) High per capita income
(b) High economic growth.
(c) Equal distribution of income
(d) Full employment
(e) Stable prices

Comparisons between economic systems are usually done by analyzing empirical data for countries that are selected because they are assumed to be typical representatives for the system one wants to study. An example of this is Bergson (1971). Other methods suggested are enumerations and evaluations of the savings that would arise if a country went from one economic system to another. Baran, Sweezy (1966), computes various savings (no advertising, no planned obsoloscence of products, etc.) which would be the result if socialism should replace capitalism.

Other methods use an analysis of models describing the systems and try to evaluate the systems analytically. Examples of such studies are Lancaster (1973) and Näslund (1973).

In this chapter we shall follow Nuti (1970) and show how double-switching (see Chapter 4) and the Pasinetti-theorem (see Chapter 5) can be used in comparing of a capitalistic and socialistic system with respect to objective (a) and, to some extent, (b) in the table above.

We should ideally begin by giving a rather complete definition of what we mean by a capitalistic and socialistic economic system. We shall, however, only mention a few aspects and refer the reader to Grossman (1967) and Hahn (1965) for more complete presentations.

With a capitalistic system we mean an economy where the enterprises are owned by private citizens (the capitalists) who employ workers from a labor market. The capitalists can borrow any amount of money at a fixed rate of interest, r. When the wages are paid, the surplus goes to the capitalist, who consumes one part and invests the rest. When selecting between techniques of production the one which maximizes the value of the assets of the capitalist at the prevailing rate of interest is chosen.

Under socialism the state owns the means of production, determines the wage rate and distributes resources to the administrative units. The state determines the method of production to be used and selects the one which gives the highest possible consumption per person at the rate of growth which is consistent with full employment.

10.1 "Double switching" and efficiency

We shall illustrate how double switching can be used in comparing a socialistic and capitalistic economy. This we shall do using an artificial economy, where only one good is produced which can be used both as a consumption good and to manufacture machines.

To produce a new machine other, previously produced, machines are used together with labor. The amount of machines necessary in period i to produce a machine we denote by $-a_i$. After a certain number of periods k we assume that the machine is completed and that in following periods, e.g. j, gives a surplus a_j machines.

The present value of investment in the machine is

$$v = \sum_{i=0}^{n} (a_i - wl_i)(1+r)^{-i} \tag{10.1}$$

where

l_i = number of workers in period i
w = wage rate
r = rate of interest

We assume that the production described above takes place both in a capitalistic and a socialistic economy.

At full employment there will be competition in the labour market which means that the present value, v, can be forced down to zero. When v=0 we can solve w from (10.1)

$$w = \frac{\sum_{i=0}^{n} a_i (1+r)^{-i}}{\sum_{i=0}^{n} l_i (1+r)^{-i}} \tag{10.2}$$

Equation (10.2) is the factor-price-frontier which we discussed in chapter 4.

If we now have many techniques to choose from,double switching is possible as described before. Two techniques A and B are equally good at a rate of interest r such that (where a_{A_i} and l_{A_i} refer to technique A etc)

$$\frac{\sum_{i=0}^{n_A} a_{A_i} (1+r)^{-i}}{\sum_{i=0}^{n_A} l_{A_i} (1+r)^{-i}} = \frac{\sum_{i=1}^{n_B} a_{B_i} (1+r)^{-i}}{\sum_{i=1}^{n_B} l_{B_i} (1+r)^{-i}} \tag{10.3}$$

If (10.3) has more than one positive solution we will have reswitching

Assume that L_t denotes the number of people working on machines which society starts to produce at time t. The machines, the production of which started year t-i, are now i years old and as was mentioned before a machine which is i years old uses l_i workers and a_i units of machines. The total number of workers who work on machines, the production of which started at t-i, is (where g is the growth rate)

$$L_{t-i} = L_t(1+g)^{-i}\, l_i \qquad (10.4)$$

The total number of persons employed at t is

$$N_t = \sum_{i=0}^{n} L_t(1+g)^{-i}\, L_t \qquad (10.5)$$

Total consumption of the good produced is

$$C_t = \sum_{i=0}^{n} L_t(1+g)^{-i}\, a_i \qquad (10.6)$$

(Note that this is the difference between what is invested during k periods and produced during n-(k+1) periods.)

If we divide (10.6) by (10.5) we obtain consumption per capita

$$c = \frac{\sum_{i=0}^{n} a_i(1+g)^{-i}}{\sum_{i=0}^{n} l_i(1+g)^{-i}} \qquad (10.7)$$

We now observe that expressions (10.2) and (10.7) look alike except for the fact that (10.2) is a function of r and c is a function of g. We can represent them in the same figure with w and c along the y-axis and g and r along the x-axis. The two curves a and b indicate two different techniques.

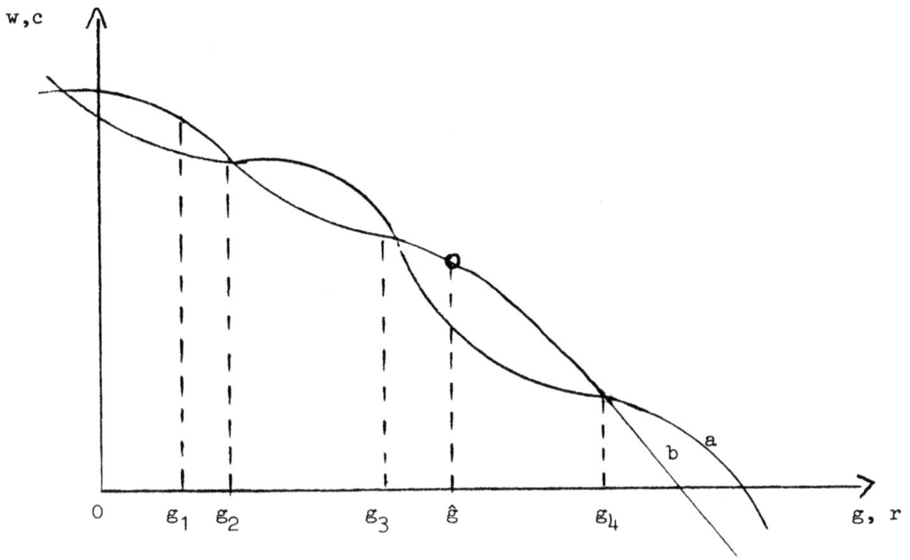

Figure 1

From Figure 1 we can see that the same conclusions can be made regarding how consumption varies with growth as we found regarding the relation between the wage rate and the rate of interest. If a certain technique is the best one at a rate of growth $\bar{g}$ it may also be best at another g_1, see Figure 1. This conclusion has implications for the present concern over environment and energy.

If Sweden selects a lower rate of growth (goes from $\bar{g}$ to g_1), then this means that we return to the same technique with the same consequences for environment and energy. Therefore one cannot be sure that a reduction of growth will lead to a less capital intensive technique using less energy. (The reader must however observe that we here study a situation when the population is growing and we try to find the best possible constant consumption per capita at different rates of growth of the population and production,)

From Figure 1 we can see that if e.g. $r=\hat{g}$, then the planned economy and the capitalistic will select the same technique. If the rate of interest rate lies between g_3 and g_4 we see that the capitalistic economy selects the same technique and this might happen also when the rate of interest differs much from $\hat{g}$, e.g. at g_1. In Figure 1 we have shown a situation when a rate of interest between 0-g, induces a capitalistic system to select the same technique as a socialistic economy at the rate of growth $\hat{g}$. (In both situations technique b will be selected.)

From this it follows that the difference between $\hat{g}$ and r cannot serve as a measure of lacking efficiency and also that one cannot in general say if the rate of interest ought to be increased or decreased in order for an optimal technique to be selected in a capitalistic economy.

From the discussion so far it follows that if we assume that the socialistic system can find an optimum, it is possible that a capitalistic system is less efficient and it is difficult to know how one shall use economic policy to make the capitalistic system more efficient.

10.2 Macroeconomic equilibrium

Except the problem of selecting the best technique within the existing economic system one has the problem of maintaining economic stability.

Under socialism the workers will pay taxes which is a fraction, b, of their salary and we assume that they save a fraction s_w of the net salary. If the public consumption is z we can write total consumption as follows.

$$z+(1-s_w)(1-b)w = c \tag{10.8}$$

Regardless how the workers invest their saving in a socialistic economy there probably will exist limitations regarding these investment such that the workers do not become "capitalists") the planner can always adjust z and b such that (10.8) holds.

In a capitalistic system the macro-economic equilibrium is achieved in a different way. The profit and how profit is measured plays an important role. The capitalists receive the difference between what is produced and what is paid as salary.

We know from Chapter 5 that Pasinetti has shown that if the workers save a fraction $s_w>0$ of their salary and if this gives them control over that share of profits which corresponds to their fraction of saving, then the workers' propensity to save does not influence the determination of capitals share of total income.

Let us assume that the workers propensity to save is zero.

If the fraction of profit which is saved is s_p, we obtain (y is income), then we can write total consumption c(r,g) as follows

$$w(r) + (1-s_p)(y(r,g)-w(r)) \tag{10.9}$$

From equation (10.9) we can solve for s_p

$$s_p = \frac{y(r,g)-c(r,g)}{y(r,g)-w(r)} \tag{10.10}$$

We know from the previous discussion in this chapter that consumption depends both of the rate of growth, g, and the rate of interest, r.

We know that consumption is determined by (see 10.7)

$$c(g,r,) = \frac{\sum_{i=0}^{n} a_i(1+g)^{-i}}{\sum_{i=0}^{n} l_i(1+g)^{-i}} \tag{10.11}$$

i.e. consumption is determined by the rate of growth and a_i and l_i. We saw in (10.3) that the technique which is selected depends upon the existing rate of interest which means that a_i and l_i depend upon the rate of interest.

If g=r, i.e. if the economy enjoys stationary growth according to the Golden Rule, then we find from (10.10) that $s_p=1$ must hold. (Since the expression for w and c, (10.2) and (10.7) become identical.)

We have here illustrated two ways according to which a capitalistic and socialistic economy differ by using results from two earlier chapters, namely Chapters 4 and 5.

The doubleswitching discussed in Chapter 4 can also be shown to apply to the choice of techniques at different rates of growth (in Chapter 4 the problem was to select technique at different rates of interest). The result is that a capitalistic economy which is governed by variations of the rate of interest can select "wrong" technique at the existing rate of growth, i.e. a technique which does not maximize long run consumption per capita.

By using the results in Chapter 5 regarding macroeconomc equilibrium growth we have shown that it is not possible to choose growth according to the Golden Rule as soon as the capitalists consume any part of their income, i.e. as soon as $s<1$.

It is naturally of interest to try to compare empirically how effective socialistic and capitalistic systems are and try, at least approximately, "test" theories of the kind described above. As was mentioned above empirical comparisons have been made realting goods and services produced to the production factors used to manufacture them. Such indices have been constructed by Bergson (1971) and used to evaluate the efficiency of capitalistic and socialistic economies, see Table 1.

National income per unit of factor inpact in different countries 1960		
	Not adjusted for labor quality. Based upon U.S. prices	Adjusted for labor quality. Based upon U.S. prices
USA	100	100
Northwester Europe	63	67
Germany	63	69
Great Britain	63	66
France	62	66
Italy	45	52
USSR	45	56

<u>Table 1</u> See Bergson (1971)

From Table 1 we see that socialistic economies perform relatively poorly. In the next section we shall discuss this result and at the same time we shall comment on the frequently discussed "convergence problem", i.e. if socialistic and capitalistic systems develop towards a similar economy.

10.3 Convergence or not

In Table 1 we showed an empirical comparison between the efficiency of different economic systems. We shall not discuss the properties of the index used by Bergson (for a critical discussion, see Domar (1971)), but rather discuss some general problems associated with the comparison between economies.

Some frequently mentioned reasons why empirical comparisons of this kind often are misleading are

(a) The countries compared are at different stages of economic development. This means that we must know that the countries being compared have reached approximately the same degree of development to make a comparison meaningful. If we compare the U.S.A. with a newly independent socialist country in Africa, an index says nothing about the relative efficiency of the two economic systems.

(b) One kind of critique against measurements of G.N.P. in socialist states is that every product produced is measured by the planner, regardless of whether people want or can use the product. The Russian comic magazine Krokodil often describes airconditioners that do not cool the air and ice boxes that cannot be used because there is no ice

(c) The capitalistic economies, on the other hand, are said to exaggerate their efficiency buy including into G.N.P. luxury consumption which does not in any reasonable way correspond to human needs. "Champagne is being produced at the same time as children die due to lack of milk." (G.G. Shaw)

However, many persons now believe that the differences between socialistic and capitalistic economies are being reduced both with respect to economic development and their functioning. (For a discussion see Grossman (1967).) By using growth as an important objective one might assume that economic development will be similar, regardless of economic system. That socialistic countries strive for growth can be illustrated by Wilczynski (1972, p. 39):

"The waste and stagnation that prevailed in the early 1960's convinced (East European) economists and political leaders of the need for activating intensive sources for growth."

As the degree of industrial development grows in socialistic and capitalistic countries, they can be expected to become more similar in many respects. The f rming sector declines, urbanization and the service sector goes up, modern science is built into production, and growth is planned at the same time as one encounters pollution and energy problems as well as increased alienation in a strongly mechanized production process.

In the capitalistic countries, the state becomes more important due to increased emphasis on services such as education, health pension plans, and an accepted view that the government must take responsibility for full employment, the distribution of income, and for economic growth which leads to interventions in the market economy.

The analysis undertaken in Sections 10.1 and 10.2 assumes both that the systems had arrived at approximately the same degree of development and that there was some degree of convergence. These assumptions must be made because we allow the same techniques to be available to both systems and the only problem is to select between them.

If the systems are very different with respect to technological development and organization, the comparisons of the type shown in Sections 10.1 and 10.2 lose much of their meaning. It is only valid when economies both maintain their main socialistic and capitalistic traits and at the same time have undergone a substantial degree of convergence.

10.4 Conclusions

Comparisons of economic systems are of great practical and theoretical interest. Politically, comparisons are used to persuade states from the third world to select an efficient economic system, or as a threat against members of a different system as expressed by Nikita Chrusjtjov "We will bury you".

In this chapter we have shown how the theories which we described in Part 1 can be used in this debate. At the same time we wish to emphasize the difficulties associated with comparisons of economic systems. These difficulties, however, enchance the need for good methods of analysis.

CHAPTER 11 CONCLUDING REMARKS

Many economists would probably agree that the most important problems in economics are

(1) the determination of value,

(2) the determination of prices,

(3) analysis of factors determining the distribution of income.

There exists a fundamental difference between economists regarding how these problems ought to be analyzed. The classical economists Ricardo and Marx put the conditions and relations associated with production in the center. These conditions determined values, prices, and the distribution of income.

The revolution which is usually associated with the British economist Jevons had the effect that the emphasis in the analysis was moved from the production side to demand and the utility of the individual consumers. Values and prices are determined on the market as expressed by e.g. Walras (1874):

> "Though it is true that productive services are bought and sold in their own special markets, nevertheless the prices of these services are determined in the market for products."

These market prices are determined independently of institutional circumstances and it is e.g. impossible for the labor unions to influence prices (on labor) and therefore the distribution of income.

It is against the foundations of this economic theory, which ultimately builds upon Jevons and Walras, which the new critique is turned.

When determining prices and thereby the distribution of income, the theory must start from marginal productivity and marginal utility. It is marginal productivity which determines the price of a factor of production, but if one uses capital as a factor of production, there must exist a homogenous capital good (before prices are determined), the marginal productivity of which one can measure. This homogenous capital good should be part of a production function such that when the factor prices change, a wage increase and a falling rate of interest will cause the selection of a more capital intensive technique.

Marginal utility on the other hand depends upon the distribution of income, and it cannot then be a factor determining prices and the distribution of income.

The modern critique thus is directed against essential foundations of important parts of economic theory. We have illustrated some important aspects of the historical development, and the most important aspects of Sraffa's critique of the neo-classical theory, which e.g. implies that the distribution of income must be determined without using the production function.

We have devoted a large part of Chapter 2 to the theory developed by Ricardo since it is possible to make some "minor" modifications and construct a theory which builds upon Ricardo's basic ideas. This we have tried to describe in Chapters 3 - 5.

Ricardo needed a standard commodity which did not change its value when the distribution of income changed. Such a commodity was constructed by Sraffa as was described in Chapter 3. Ricardo needed another theory of demand (which does not build upon Say's law) and another population theory which was not the pessimistic one developed by Malthus, and he also had to consider technological development. This is brought into the theory in Chapter 5.

Perhaps the greatest importance of a different foundation of economic theory lies in the fact that the foundation determines the selection of problems and explanatory variables in the analysis of economic problems on the surface. Part II has the purpose of showing how the theory presented in Part I can be used for discussions of current problems in society. The selection of problem areas, and the way they are analyzed, is influenced by the attempt to show how the theory can be used.

We shall conclude by indicating some consequences of the fact that different theoretical foundations are used in solving economic problems.

I. In Sraffa's model it is emphasized that prices depend upon how goods are produced. Production of commodities must -- for a society to be able to survive -- reproduce those commodities which are being used in the production process (with the exception of certain natural resources).

Exchanges on the market are therefore influenced by the production process. This means that the position of an individual or a corporation on the market is primarily determined by the position the individual has in production and the power that he has to get a part of the surplus that might exist after one has used that which is necessary for the reproduction of the economic system.

In neo-classical theory prices are determined by exchanges on a market which is not directly influenced by the way in which the production process is organized, and those acting on the market are only characterized by their being buyers or sellers.

II. This difference regarding price formation influences how one views the functioning of the market and its way to operate. In the neo-classical representation everybody gains by exchanges on the market. In Sraffa's presentation as well as in Ricardo's and Marx', one partly tend to win at the expense of somebody else.

III. The neo-classical theory emphasizes that, in equilibrium on the market, different factors of production get their value determined by their marginal productivity. The presentation in Chapter 4 has shown that this is a doubtful method with respect to the value of capital.

These ways of viewing the role of the market, the role of the exchange process, and the contributions and rewards given to the individuals participating, are examples of fundamental differences, which affect the problems one selects to study and which variables and relations one considers to be important in the analysis.

We hope that the book has stimulated the reader in an ongoing search for the foundation he prefers to use.

REFERENCES

ALTHUSSER, L. and BALIBAR, F., 1970, Reading Capital. London

ANDERSSON, R. and MEIDNER, R., 1973, Arbetsmarknadspolitik och stabilisering. Vänersborg

ARROW, K.J., 1962, The Economic Implications of Learning by Doing. Review of Economic Studies, Vol. 29, 155-73

BARAN, P.A. and SWEEZY, P.M., 1966, Monopoly Capital. Pelican Book

BERGSON, A., 1968, Planning and Productivity under Soviet Socialism. New York

BERGSON, A., 1971, Comparative Productivity and Efficiency in the Soviet Union and the United States. In Eckstein (ed.), Comparison of Economic Systems. University of California Press

BHADURI, A., 1969, On the Significance of Recent Controversies on Capital Theory: A Marxian View. Economic Journal, 532-39

BHARADWAJ, K., 1970, On the Maximum Number of Switches between Two Production Systems. Schweizerische Zeitschrift für Volkswirtschaft und Statistik, 409-29

BLISS, C.J., 1975, Capital Theory and the Distribution of Income, North-Holland Publishing Company

BOSE, A., 1964, The "labour approach" and the "commodity approach" in Mr Sraffa's price theory. Economic Journal, Vol. 74, 723-26

BOSE, A., 1965, Consumers' demand, distributive shares and prices. Economic Journal, Vol. 75, 771-86

BURSTEIN, M.L., 1969, The Index-Number Problem. In Clower, R.L. (ed.), Monetary Theory, 61-64. Middlesex

CLOWER, R.W., 1967, A reconsideration of the microfoundations of monetary theory. Western Economic Journal, Vol. 6, 1-9

DALTON, G., 1974, Economic Systems & Society - Capitalism, Communism, and the Third World. Penguin

DEBREU, G. and HERSTEIN, I.N., 1953, Non-negative square matrices. Econometrica, Vol. 21, 597-607

DENCIK, P., HERLITZ, L., and LUNDVALL, B.-Å., 1969, Marxismens politiska ekonomi - en introduktion. Kristianstad

DOBB, M., 1940, Political Economy and Capitalism: Some Essays in Economic Tradition. London

DOBB, M., 1970, The Sraffa System and Critique of the Neo-Classical Theory of Distribution. De Economist, 347-62. Also in Hunt and Schwartz, 1972, A Critique of Economic Theory. Penguin

DOBB, M., 1973, Theories of Value and Distribution since Adam Smith. Cambridge

DOMAR, E.D., 1947, Expansion and Employment. American Economic Review, March

DOMAR, E.D., 1971, On the Measurement of Comparative Efficiency. In Eckstein (ed.), Comparison of Economic Systems. University of California Press

ECKSTEIN, A., 1971, Comparison of Economic Systems. University of California Press

FISHER, I., 1907, The Rate of Interest. MacMillan

FISHER, I., 1930, The Theory of Interest. MacMillan

FOUCAULT, M., 1972, Vetandets arkeologi. VEB Druckhaus. Köthen DDR

GAREGAGNI, P., 1970, Heterogeneous Capital, the Production Function and the Theory of Distribution. Review of Economic Studies, pp. 407-36. (Also in Hunt, E.K. and Schwartz, J.G., 1972, A Critique of Economic Theory. Penguin.)

GROSSMAN, G., 1967, Economic Systems. Prentice Hall

HAHN, F.H. and MATTHEWS, R.C.O., 1970, Growth and Technical Progress: A Survey. In Sen, A. (ed.), Growth Economics, pp. 367-97. Middlesex

HALN, G.N., 1970, Economic Systems: A Comparative Analysis. Holt, Rinehart and Winston

HARCOURT, G.C., 1972, Some Cambridge Controversies in the Theory of Capital. Cambridge

HARCOURT, G.C. and MASSARO, V.G., 1964, A Note on Mr Sraffa's Sub-systems. Economic Journal, Vol. 74, pp. 715-22

HUNT, E.K. and SCHWARTZ, J.G., 1972, A Critique of Economic Theory. Penguin

JOHANSEN, L., 1963, Labour Theory of Value and Marginal Utilities. Economics of Planning, Vol. 3, pp. 89-103

KALDOR, N., 1955-56, Alternative Theories of Distribution. Review of Economic Studies, Vol. 23, pp. 83-100

KALDOR, N., 1966, Marginal Productivity and Macroeconomic Theories of Distribution. Review of Economic Studies, Vol. 33, pp. 309-19

KALDOR, N., 1970, Some Fallacies in the Interpretation of Kaldor. Review of Economic Studies, Vol. 37, pp. 1-7

KALDOR, N. and MIRRLEES, J.A., 1961-62, A New Model of Economic Growth. Review of Economic Studies, Vol. 29, pp. 174-90

KENNEDY, C. and THIRLWALL, A.P., 1972, Tehcnical Progress. Economic Journal, Vol. 82, March

KLEIN, E., 1973, Mathematical Methods in Theoretical Economics. New York

LANCASTER, K., 1973, The Dynamic Inefficiency of Capitalism. Journal of Political Economy, September

MACHLUP, F., 1960, Another View of Cost-push and Demand-pull Inflation. Review of Economics and Statistics, Vol. 42, pp. 125-39

MACHLUP, F., 1962, The Supply of Inventros and Inventions. In The Rate and Direction of Inventive Activity. Princeton

MEADE, J.E., 1966, The Outcome of the Pasinetti-Process. Economic Journal, pp. 161-65

MEDIO, A., 1972, Profits and Surplus-Value: Appearance and Reality in Capitalist Production. In Hunt, E.K. and Schwartz, J.G. (eds.). A Critique of Economic Theory, pp. 312-316. Middlesex

MEEK, R.L., 1974, Value in the History of Economic Thought. History of Political Economy, Vol. 6, pp. 246-60

MEIDNER, R. et al., 1975, Löntagarfonder

MORISHIMA, M., 1973, Marx's Economics. Cambridge

MYRDAL, G., 1972, Vetenskap och politik i nationalekonomi. Stockholm

MYRDAL, G., 1973, I stället för memoarer; kritiska essäer om nationalekonomin. Vänersborg

NELL, E.J., 1967, Theories of Growth and Theories of Value. Economic Development and Cultural Change, Vol. 16, pp. 15-26

NELL, E.J., 1974, Egendom och produktionsmedel. En inledning till Cambridge-kontroversen. Nordisk Tidskrift för Politisk Ekonomi, Nr 1, pp. 7-22

NEWMAN, P., 1962, Production of Commodities by Means of Commodities. Schweizerische Zeitung für Volkswirtschaft und Statistik, pp. 58-75

NUTI, D.M., 1969, The Degree of Monopoly in the Kaldor-Mirrlees Growth Model. Review of Economic Studies, Vol. 36, pp. 257-59

NUTI, D.M., 1970, Capitalism, Socialism and Steady Growth. Economic Journal 80, pp. 32-57

NUTI, D.M., 1970, Vulgar Economy. In The Theory of Distribution, De Economist, pp. 363-69. (Also in Hunt, E.K. and Schwartz, J.G., 1972, A Critique of Economic Theory. Penguin.)

NÄSLUND, B., 1973, Labor Power and Income Distribution. The Swedish Journal of Economics, Vol. 73, pp. 128-42

OWEN, R., 1969, Report to the County of Lanark, Middlesex. (1:st ed. 1820)

PAISH, F.W., 1969, Unemployment and Price Stability. In Ball, R.J. and Doyle, P. (eds.), Inflation, pp. 219-54. Middlesex

PASINETTI, L., 1960, A Mathematical Formulation of the Ricardian System. Review of Economic Studies, Vol. 27, pp. 78-98

PASINETTI, L., 1969, Switches of Technique and the "Rate of Return" in Capital Theory. Economic Journal, pp. 508-31

PASINETTI, L., 1974, Growth and Income Distribution. Cambridge

PATINKIN, D., 1965, Money, Interst and Prices. New York

RICARDO, D., 1971, Principles of Political Economy and Taxation. Middlesex. (1:st ed. 1817)

ROBINSON, J., 1961, Prelude to a Critique of Economic Theory. Oxford Economic Papers, pp. 7-14. (Also in Hunt, E.K. and Schwartz, J.G., 1972, A Critique of Economic Theory. Penguin.)

ROBINSON, J., 1969, A Further Note. Review of Economic Studies, Vol. 36, pp. 260-62

ROBINSON, J., 1973, Collected Economic Papers, Volume IV. Oxford

ROBINSON, J. and EATWELL, J., 1973, An Introduction to Modern Economics. London

ROBINSON, J. and NAQVI, K.A., 1967, The Badly Behaved Production Function. Quarterly Journal of Economics, pp. 579-91

RYMES, T.K., 1971, On Concepts of Capital and Technical Change. Cambridge

SAMUELSON, P.A., 1962, Parable and Realism in Capital Theory: the Surrogate Production Function. Review of Economic Studies, pp. 193-206

SAMUELSON, P.A., 1966, A Summing Up. Quarterly Journal of Economics, pp. 568-83

SAMUELSON, P.A. and MODIGLIANI, F., 1966, The Pasinetti Paradox in Neo-Classical and More General Models. Review of Economic Studies, pp. 269-301

SCHULTZ, C.L., 1969, Sectoral Shifts and Inflation. In Ball, R.J. and Doyle, P. (eds.), Inflation, pp. 209-18. Middlesex

SELLSTEDT, B., 1972, Selection of Product Development Projects under Uncertainty. Stockholm

SOLOW, R.M., 1956, A Contribution to the Theory of Economic Growth. Quarterly Journal of Economics, pp. 65-94.

SOLOW, R.M., 1963, Capital Theory and the Rate of Return. North-Holland

SOLOW, R.M., 1967, The Interest Rate and Transition between Techniques. In Feinstein, C.H. (ed.), Socialism, Capitalism and Economic Growth. Essays presented to Maurice Dobb, pp. 30-39. Cambridge University Press

SOMMERVILLE, J.M., 1972, Marx' och Engels' samhällsvetenskapliga metod. Finland

SRAFFA, P., 1960, Production of Commodities by Means of Commodities. Cambridge

STREETEN, P., 1969, Productivity Inflation. In Ball, R.J. and Doyle, P. (eds.), Inflation. Middlesex

WILCZYNSKI, J., 1972, Socialist Economic Development and Reform. MacMillan

DE VYLDER, S., 1970, Vinsten hos klassikerna och Marx. Häften för kritiska studier, Nr 7-8

YEAGER, L.B., 1968, Essential Properties of the Medium of Exchange. Kyklos, Vol. 21, pp. 45-68

Vol. 115: Anderson et al., Foundations of System Theory: Finitary and Infinitary Conditions. VII, 93 pages. 1976

Vol. 116: K. Miyazawa, Input-Output Analysis and the Structure of Income Distribution. IX, 135 pages. 1976.

Vol. 117: Optimization and Operations Research. Proceedings 1975. Edited by W. Oettli and K. Ritter. IV, 316 pages. 1976.

Vol. 118: Traffic Equilibrium Methods, Proceedings 1974. Edited by M. A. Florian. XXIII, 432 pages. 1976.

Vol. 119: Inflation in Small Countries. Proceedings 1974. Edited by H. Frisch. VI, 356 pages. 1976.

Vol. 120: G. Hasenkamp, Specification and Estimation of Multiple-Output Production Functions. VII, 151 pages. 1976.

Vol. 121: J. W. Cohen, On Regenerative Processes in Queueing Theory. IX, 93 pages. 1976.

Vol. 122: M. S. Bazaraa, and C. M. Shetty,Foundations of Optimization VI. 193 pages. 1976

Vol. 123: Multiple Criteria Decision Making. Kyoto 1975. Edited by M. Zeleny. XXVII, 345 pages. 1976.

Vol. 124: M. J. Todd. The Computation of Fixed Points and Applications. VII, 129 pages. 1976.

Vol. 125: Karl C. Mosler. Optimale Transportnetze. Zur Bestimmung ihres kostengünstigsten Standorts bei gegebener Nachfrage. VI, 142 Seiten. 1976.

Vol. 126: Energy, Regional Science and Public Policy. Energy and Environment I. Proceedings 1975. Edited by M. Chatterji and P. Van Rompuy. VIII, 316 pages. 1976.

Vol. 127: Environment, Regional Science and Interregional Modeling. Energy and Environment II. Proceedings 1975. Edited by M. Chatterji and P. Van Rompuy. IX, 211 pages. 1976.

Vol. 128: Integer Programming and Related Areas. A Classified Bibliography. Edited by C. Kastning. XII, 495 pages. 1976.

Vol. 129: H.-J. Lüthi, Komplementaritäts- und Fixpunktalgorithmen in der mathematischen Programmierung. Spieltheorie und Ökonomie. VII, 145 Seiten. 1976.

Vol. 130: Multiple Criteria Decision Making, Jouy-en-Josas, France. Proceedings 1975. Edited by H. Thiriez and S. Zionts. VI, 409 pages. 1976.

Vol. 131: Mathematical Systems Theory. Proceedings 1975. Edited by G. Marchesini and S. K. Mitter. X, 408 pages. 1976.

Vol. 132: U. H. Funke, Mathematical Models in Marketing. A Collection of Abstracts. XX, 514 pages. 1976.

Vol. 133: Warsaw Fall Seminars in Mathematical Economics 1975. Edited by M. W. Loś, J. Loś, and A. Wieczorek. V. 159 pages. 1976.

Vol. 134: Computing Methods in Applied Sciences and Engineering. Proceedings 1975. VIII, 390 pages. 1976.

Vol. 135: H. Haga, A Disequilibrium – Equilibrium Model with Money and Bonds. A Keynesian – Walrasian Synthesis. VI, 119 pages. 1976.

Vol. 136: E. Kofler und G. Menges, Entscheidungen bei unvollständiger Information. XII, 357 Seiten. 1976.

Vol. 137: R. Wets, Grundlagen Konvexer Optimierung. VI, 146 Seiten. 1976.

Vol. 138: K. Okuguchi, Expectations and Stability in Oligopoly Models. VI, 103 pages. 1976.

Vol. 139: Production Theory and Its Applications. Proceedings. Edited by H. Albach and G. Bergendahl. VIII, 193 pages. 1977.

Vol. 140: W. Eichhorn and J. Voeller, Theory of the Price Index. Fisher's Test Approach and Generalizations. VII, 95 pages. 1976.

Vol. 141: Mathematical Economics and Game Theory. Essays in Honor of Oskar Morgenstern. Edited by R. Henn and O. Moeschlin. XIV, 703 pages. 1977.

Vol. 142: J. S. Lane, On Optimal Population Paths. V, 123 pages. 1977.

Vol. 143: B. Näslund, An Analysis of Economic Size Distributions. XV, 100 pages. 1977.

Vol. 144: Convex Analysis and Its Applications. Proceedings 1976. Edited by A. Auslender. VI, 219 pages. 1977.

Vol. 145: J. Rosenmüller, Extreme Games and Their Solutions. IV, 126 pages. 1977.

Vol. 146: In Search of Economic Indicators. Edited by W. H. Strigel. XVI, 198 pages. 1977.

Vol. 147: Resource Allocation and Division of Space. Proceedings. Edited by T. Fujii and R. Sato. VIII, 184 pages. 1977.

Vol. 148: C. E. Mandl, Simulationstechnik und Simulationsmodelle in den Sozial- und Wirtschaftswissenschaften. IX, 173 Seiten. 1977.

Vol. 149: Stationäre und schrumpfende Bevölkerungen: Demographisches Null- und Negativwachstum in Österreich. Herausgegeben von G. Feichtinger. VI, 262 Seiten. 1977.

Vol. 150: Bauer et al., Supercritical Wing Sections III. VI, 179 pages. 1977.

Vol. 151: C. A. Schneeweiß, Inventory-Production Theory. VI, 116 pages. 1977.

Vol. 152: Kirsch et al., Notwendige Optimalitätsbedingungen und ihre Anwendung. VI, 157 Seiten. 1978.

Vol. 153: Kombinatorische Entscheidungsprobleme: Methoden und Anwendungen. Herausgegeben von T. M. Liebling und M. Rössler. VIII, 206 Seiten. 1978.

Vol. 154: Problems and Instruments of Business Cycle Analysis. Proceedings 1977. Edited by W. H. Strigel. VI, 442 pages. 1978.

Vol. 155: Multiple Criteria Problem Solving. Proceedings 1977. Edited by S. Zionts. VIII, 567 pages. 1978.

Vol. 156: B. Näslund and B. Sellstedt, Neo-Ricardian Theory. With Applications to Some Current Economic Problems. VI, 165 pages. 1978.

Ökonometrie und Unternehmensforschung
Econometrics and Operations Research

Vol. I	Nichtlineare Programmierung. Von H. P. Künzi und W. Krelle unter Mitwirkung von W. Oettli. Vergriffen
Vol. II	Lineare Programmierung und Erweiterungen. Von G. B. Dantzig. Ins Deutsche übertragen und bearbeitet von A. Jaeger. – Mit 103 Abbildungen. XVI, 712 Seiten. 1966. Geb.
Vol. III	Stochastic Processes. By M. Girault. – With 35 figures. XII, 126 pages. 1966. Cloth
Vol. IV	Methoden der Unternehmensforschung im Versicherungswesen. Von K. H. Wolff. – Mit 14 Diagrammen. VIII, 266 Seiten. 1966. Geb.
Vol. V	The Theory of Max-Min and its Application to Weapons Allocation Problems. By John M. Danskin. – With 6 figures. X, 126 pages. 1967. Cloth
Vol. VI	Entscheidungskriterien bei Risiko. Von H. Schneeweiss. – Mit 35 Abbildungen. XII, 215 Seiten. 1967. Geb.
Vol. VII	Boolean Methods in Operations Research and Related Areas. By P. L. Hammer (Ivănescu) and S. Rudeanu. With a preface by R. Bellman. – With 25 figures. XVI, 329 pages. 1968. Cloth
Vol. VIII	Strategy for R & D: Studies in the Microeconomics of Development. By Th. Marschak. Th K. Glennan Jr., and R. Summers. – With 44 figures. XIV, 330 pages. 1967. Cloth
Vol. IX	Dynamic Programming of Economic Decisions. By M. J. Beckmann. – With 9 figures XII, 143 pages. 1968. Cloth
Vol. X	Input-Output-Analyse. Von J. Schumann. – Mit 12 Abbildungen. X, 311 Seiten. 1968. Geb.
Vol. XI	Produktionstheorie. Von W. Wittmann. – Mit 54 Abbildungen. VIII, 177 Seiten. 1968. Geb.
Vol. XII	Sensitivitätsanalysen und parametrische Programmierung. Von W. Dinkelbach. – Mit 20 Abbildungen. XI, 190 Seiten. 1969. Geb.
Vol. XIII	Graphentheoretische Methoden und ihre Anwendungen. Von W. Knödel. – Mit 24 Abbildungen. VIII, 111 Seiten. 1969. Geb.
Vol. XIV	Praktische Studien zur Unternehmensforschung. Von E. Nievergelt, O. Müller, F. E. Schlaepfer und W. H. Landis. – Mit 82 Abbildungen. XII, 240 Seiten. 1970. Geb.
Vol. XV	Optimale Reihenfolgen. Von H. Müller-Merbach. – Mit 45 Abbildungen. IX, 225 Seiten. 1970. Geb.
Vol. XVI	Preispolitik der Mehrproduktenunternehmung in der statischen Theorie. Von R. Selten. – Mit 20 Abbildungen. VIII, 195 Seiten. 1970. Geb.
Vol. XVII	Information Theory for Systems Engineers. By L. P. Hyvärinen. – With 42 figures. VIII, 197 pages. 1970. Cloth
Vol. XVIII	Unternehmensforschung im Bergbau. Von F. L. Wilke. – Mit 29 Abbildungen. VIII, 150 Seiten. 1972. Geb.
Vol. XIX	Anti-Äquilibrium. Von J. Kornai. – Mit 31 Abbildungen. XVI, 382 Seiten. 1975. Geb.
Vol. XX	Mathematische Optimierung. Von E. Blum, W. Oettli. IX, 413 Seiten. (davon 75 Seiten Bibliographie). 1975. Geb.
Vol. XXI	Stochastic Linear Programming. By P. Kall. VI, 95 pages. 1976. Cloth.

GPSR Compliance

The European Union's (EU) General Product Safety Regulation (GPSR) is a set of rules that requires consumer products to be safe and our obligations to ensure this.

If you have any concerns about our products, you can contact us on ProductSafety@springernature.com

In case Publisher is established outside the EU, the EU authorized representative is:

Springer Nature Customer Service Center GmbH
Europaplatz 3
69115 Heidelberg, Germany

Batch number: 10844827

Printed by Printforce, the Netherlands